# Social-Environmental Conflicts, Extractivism and Human Rights in Latin America

This book focuses on the issues of global environmental injustice and human rights violations and explores the scope and limits of the potential of human rights to influence environmental justice. It offers a multidisciplinary perspective on contemporary development discussions, analysing some of the crucial challenges, contradictions and promises within current environmental and human rights practices in Latin America. The contributors examine how the extraction and exploitation of natural resources and the further commodification of nature have affected local communities in the region, and how these policies have impacted on the promotion and protection of human rights as communities struggle to defend their rights and territories. The book analyses the emergence of transnational activism in the context of collective action organised around socio-environmental conflicts, the infringement of basic human rights and the emergence of alternative and sometimes conflicting development models. Furthermore, it critically discusses why governments are often willing to override their commitments to sustainability and human rights to promote their development agenda.

The chapters in this book were originally published as a special issue in *The International Journal of Human Rights*.

**Malayna Raftopoulos** is an assistant professor in Latin American studies at Aalborg University, Denmark. She is also the co-editor of *Provincialising Nature: Multidisciplinary Approaches to the Politics of the Environment in Latin America* (Institute of Latin American Studies [ILAS], University of London).

# Social-Environmental Conflicts, Extractivism and Human Rights in Latin America

*Edited by*
Malayna Raftopoulos

Routledge
Taylor & Francis Group

LONDON AND NEW YORK

First published 2018 by Routledge

2 Park Square, Milton Park, Abingdon, Oxfordshire OX14 4RN
52 Vanderbilt Avenue, New York, NY 10017

*Routledge is an imprint of the Taylor & Francis Group, an informa business*

First issued in paperback 2019

*British Library Cataloguing in Publication Data*
A catalogue record for this book is available from the British Library

ISBN13: 978-0-8153-5372-0 (hbk)
ISBN13: 978-0-367-89310-1 (pbk)

Typeset in TimesNewRomanPS
by diacriTech, chennai

**Publisher's Note**
The publisher accepts responsibility for any inconsistencies that may have arisen during the conversion of this book from journal articles to book chapters, namely the possible inclusion of journal terminology.

**Disclaimer**
Every effort has been made to contact copyright holders for their permission to reprint material in this book. The publishers would be grateful to hear from any copyright holder who is not here acknowledged and will undertake to rectify any errors or omissions in future editions of this book.

# Contents

# Citation Information

The chapters in this book were originally published in *The International Journal of Human Rights*, volume 21, issue 4 (May 2017). When citing this material, please use the original page numbering for each article, as follows:

**Chapter 1**

*Contemporary debates on social-environmental conflicts, extractivism and human rights in Latin America*
Malayna Raftopoulos
*The International Journal of Human Rights*, volume 21, issue 4 (May 2017) pp. 387–404

**Chapter 2**

*'...Beggars sitting on a sack of gold': Oil exploration in the Ecuadorian Amazon as* buen vivir *and sustainable development*
Joanna Morley
*The International Journal of Human Rights*, volume 21, issue 4 (May 2017) pp. 405–441

**Chapter 3**

*State-led extractivism and the frustration of indigenous self-determined development: lessons from Bolivia*
Radosław Powęska
*The International Journal of Human Rights*, volume 21, issue 4 (May 2017) pp. 442–463

**Chapter 4**

*Ethnic rights and the dilemma of extractive development in plurinational Bolivia*
Rickard Lalander
*The International Journal of Human Rights*, volume 21, issue 4 (May 2017) pp. 464–481

**Chapter 5**

*The international human rights discourse as a strategic focus in socio-environmental conflicts: the case of hydro-electric dams in Brazil*
Marieke Riethof
*The International Journal of Human Rights*, volume 21, issue 4 (May 2017) pp. 482–499

**Chapter 6**

*Extracting justice? Colombia's commitment to mining and energy as a foundation for peace*

John-Andrew McNeish

*The International Journal of Human Rights*, volume 21, issue 4 (May 2017) pp. 500–516

For any permission-related enquiries please visit:
http://www.tandfonline.com/page/help/permissions

# Notes on Contributors

**Rickard Lalander** holds a PhD in Latin American studies from the University of Helsinki, Finland. Currently, he works as a lecturer in Development and Environmental Studies at Södertön University.

**John-Andrew McNeish** is Professor of International Environment and Development Studies at the Norwegian University of Life Sciences (NMBU), Norway.

**Joanna Morley** completed an MA in Understanding and Securing Human Rights in 2015. Her published articles include 'Extreme energy, "fracking" and human rights: a new field for human rights impact assessments?' (co-author) in the *International Journal of Human Rights*, and a book review of 'Human Rights From Community: A Rights-Based Approach to Development' by Oche Onazi in the *International Community Law Review*. Her research interests include the human rights impacts on local communities of international development policies linked to natural resource exploitation, particularly focussing on Chinese investments in Latin America and Africa.

**Radosław Powęska** is a Latin Americanist working in the fields of politics and political sociology and anthropology. He received a PhD from the University of Liverpool, UK. He is professionally associated with the Center for Latin American Studies (CESLA), University in Warsaw, Poland. His Latin American research interests include ethnopolitics, social movements, indigenous rights and state–indigenous relations, political cultures and political innovations.

**Malayna Raftopoulos** is an assistant professor in Latin American studies at Aalborg University, Denmark. She is also the co-editor of *Provincialising Nature: Multidisciplinary Approaches to the Politics of the Environment in Latin America* (Institute of Latin American Studies [ILAS], University of London Press).

**Marieke Riethof** is a lecturer in Latin American politics at the University of Liverpool, UK. Her past research and publications focused on political strategies of the labour movement in Brazil, including the Latin American regional context.

# Contemporary debates on social-environmental conflicts, extractivism and human rights in Latin America

Malayna Raftopoulos ⓘ

This opening contribution to 'Social-Environmental Conflicts, Extractivism and Human Rights in Latin America' analyses how human rights have emerged as a weapon in the political battleground over the environment as natural resource extraction has become an increasingly contested and politicised form of development. It examines the link between human rights abuses and extractivism, arguing that this new cycle of protests has opened up new political spaces for human rights based resistance. Furthermore, the explosion of socio-environmental conflicts that have accompanied the expansion and politicisation of natural resources has highlighted the different conceptualisations of nature, development and human rights that exist within Latin America. While new human rights perspectives are emerging in the region, mainstream human rights discourses are providing social movements and activists with the legal power to challenge extractivism and critique the current development agenda. However, while the application of human rights discourses can put pressure on governments, it has yielded limited concrete results largely because the state as a guardian of human rights remains fragile in Latin America and is willing to override their commitment to human and environmental rights in the pursuit of development. Lastly, individual contributions to the volume are introduced and future directions for research in natural resource development and human rights are suggested.

## Introduction

Natural resource exploitation, and the increasing number of large-scale and mega-development projects in the region, has made Latin America one of the most dangerous places for human rights activists and environmentalists in the world. Even progressive governments such as Ecuador have employed a zero-tolerance policy towards anyone opposing natural resource extraction. Ecuadorian authorities have led a campaign to vilify and stigmatise indigenous groups and social movements, labelling them 'environmental extremists' or 'terrorists' in an attempt to build a framework of acceptance for curtailing human rights in the name of development. President Rafael Correa even attempted to close down the country's leading grassroots environmental organisation, Acción Ecológica, in a clear reprisal over their support for the Shuar Indigenous People who are embroiled in a bitter conflict with the government over a planned mega-copper mine on their ancestral lands in the southern Ecuadorian Amazon. The link between human rights abuses and natural resources

1

has become the focus of growing concern as governments throughout the region push through major development projects without integrating economic, social and cultural rights. Although the continent has a long history of extracting and exploiting natural resources dating back to the colonial era, there has been a marked increase in these activities in the region in the last decade or so, associated with the strong international demand for raw materials and a cycle of high prices. However, the recent downturn in the price of minerals and hydrocarbons has further exacerbated the problem, as the decline in profits is offset by the further expansion of extractive frontiers. The proliferation of extractivist activities and its diversification into new areas such as hydroelectricity has significantly impacted on the enjoyment of human rights across the hemisphere and has become a permanent cause of social-environmental conflicts. While governments and multinational corporations have been riding the wave of the commodities boom, indigenous and peasant communities have found themselves 'at the leading edge of both the extractive capital frontier and the related social conflict'.[1] These social-environmental conflicts are not isolated but are occurring throughout the continent, engaging communities in a continual battle against natural resource exploitation and the forces of global capital, resulting in repeated and widespread clashes, violence, repression and human rights abuses perpetuated by the state or security forces. As the anthropologist Philippe Descola observed, '[o]ne does not have to be a great seer to predict that the relationship between humans and nature will, in all probability, be the most important question of the present century'.[2] Yet, despite this, the relationship between human rights, extractivism and the environment remains under-researched and under-theorised.

Human rights have emerged as a weapon in the political battleground over the environment as natural resource extraction has become an increasingly contested and politicised form of development. Latin American governments have pursued extraction relentlessly, regardless of the socio-environmental costs and the abrogation of the most fundamental human rights that this development model entails. A report published by Inter-American Commission on Human Rights in 2015, while stating that states have the freedom to exploit their natural resources through concessions and private or public investments of either a national or international nature, also importantly emphasised that these activities should not be executed at the expense of human rights and justice.[3] Along with this increasing recognition of the linkage between human rights and extractivism, questions are also being raised within human rights law over approaches to environmental protection and recognition of intercultural perspectives. The explosion of social-environmental conflicts that has accompanied the expansion of extractive activities has posed a challenge to the political and economic ideology of the current development model. This challenge comes from the new relational ontologies of local and indigenous communities and cultures who have opened up debates about the relationship between the human and non-human world, the rights of nature and human rights and duties.

While extractivism previously referred to activities that involved extracting, such as in mining, oil and gas, the term is now increasingly used to refer to the accelerated pace of natural resource exploitation at an industrial level and the construction of mega-projects and infrastructure intended to make full use of natural resources.[4] During the expansion of the extractive and infrastructure frontiers in Latin America, territories that were previously isolated or protected and 'often biologically fragile environments populated by vulnerable populations who share their land with minerals or energy sources' have been opened up for exploitation.[5] According to a study conducted by Global Witness, 2015 was the worst year on record for the murder of land and environmental defenders with a total of 185 assassinations across the globe.[6] In March 2016, the United Nations (UN)

Human Rights Council adopted a landmark resolution requiring states to ensure the rights and safety of human rights defenders working towards the realisation of economic, social and cultural rights.[7] Speaking ahead of World Environment Day 2016, The UN Special Rapporteur on Human Rights and the Environment, John Knox, along with the UN Special Rapporteur on the Situation of Human Rights Defenders, Michel Forst, and the UN Special Rapporteur on the Rights of Indigenous People, Victoria Tauli Corpuz, issued a joint statement urging governments to protect environmental rights defenders.[8] However, increasingly governments across Latin America are criminalising social protests through the use of repressive legislation, and deterring or curtailing communities and activists from political mobilisation through the use of violence, kidnapping, torture, harassment and threats. The link between environmental injustice and human rights transgressions highlights the urgent need to bring together human rights and the environment, 'two dominant legal and social discourses often assumed to have at best an uneasy, and at worse an antithetical relationship'.[9]

The murder of Berta Cáceres, a well-known activist for indigenous rights, human rights and environmental protection in Honduras in March 2016 exposed the level of violence that often accompanies mega-projects and resource extraction in Latin America as indigenous communities and governments clash over the use and control of natural resources and land. Opposing the construction of four dams designed to power future mining operations along the Gualcarque River, an area sacred to the Lenca indigenous community in western Honduras and known collectively as the Agua Zarca Dam, Cáceres waged a grassroots campaign that successfully pressured the world's largest dam developer, Sinohydro, to pull out of the project. From the Chevron case in Ecuador, to the Belo Monte dam protests in Brazil and the TIPNIS (*Territorio Indígena y Parque Nacional Isiboró Securé*) dispute in Bolivia, communities and activists across Latin America are engaged in struggles against extractive or damaging infrastructural activities taking place in their territories. Current Latin American governments' continued fidelity to the neoliberal developmental agenda, coupled with globalisation, has led to a new cycle of protests in the region and opened up new political spaces for human rights based resistance in natural resource governance to transnational networks of (indigenous) social movements, human rights actors and nongovernmental organisations to mobilise.

Despite the widespread optimism that the alternative platforms put forward by the left and centre-left governments would transcend modernist development paradigms following the legitimacy crisis of neoliberalism, apparently progressive governments have not only continued but have intensified the neoliberal policy of extractivism. This has led to a plethora of social-environmental conflicts and the continued violation of both human and environmental rights throughout Latin America as democratic processes are eroded alongside renewed efforts to expand extractive frontiers. This opening contribution examines the link between human rights abuses and extractivism, arguing that this new cycle of protests has opened up new political spaces for human rights based resistance. Furthermore, the explosion of socio-environmental conflicts that have accompanied the expansion and politicisation of natural resources has highlighted the different conceptualisations of nature, development and human rights that exist within Latin America. Peasant and indigenous communities have found themselves at the forefront of the resource wars as they clash with governments and multinational corporations over the use and control of the global commons. With current international law on environmental management by sovereign states limited to managing the environment in a manner that the misuse of natural resources does not disadvantage other states,[10] the international human rights framework has become increasingly important. The potential of human rights to act as 'language of protest' and a 'platform for change'[11] has contributed to the increasing transnationalisation of human rights discourses in the last two decades and led to the development of transnational human rights networks that bring together ordinary social actors in their pursuit against similar claims of injustice. While new human rights perspectives are emerging in the region, mainstream human rights discourses are providing social movements and activists with the legal power to challenge extractivism and critique the current development agenda. However, the application of human rights discourses so far has yielded limited results largely because the state as a guardian of human rights remains fragile in Latin America and is willing to override their commitment to human and environmental rights in the pursuit of development. The article is concluded by discussing the individual contributions to the volume and also future directions for research in natural resource development and human rights.

## Alternatives to development and extractivism

The dilemma between exploiting natural resources for socio-economic development and defending both human and environmental rights represents a major challenge for Latin American countries. Since colonial times, Latin America's relationship with natural resources has been a source of conflicting political, social and economic dynamics. As Haarstad comments, 'natural resources have traditionally been considered a curse on Latin American societies, from the plundering of the colonial era to the ills of commodity dependency in later years'.[12] Moreover, the socio-economic conditions produced by extractivist-based economies have contributed to ecological destruction, widespread poverty and social injustice throughout the region. Schmink and Jouve-Martín explain that 'Latin America's historical dependency on natural resources, both for local livelihoods and to supply an evolving global market, has made environmental issues central in policy debates and in widespread contests over the meaning and use of natural species and habitats, carried out against the region's persistent legacy of inequality'.[13] This 'curse of abundance' has created the preconditions for not just political but financial, commercial, social and energy instability.[14] This continued reliance on the exploitation of non-renewable resources perpetuates neocolonial power relations based on the export-led growth model, with incalculable environmental consequences, and further undermines democratic institutions by creating a 'paternalistic state', one 'whose political impact is a direct result of its ability to manage a higher or lower participation in the mining or oil revenues'.[15] Protecting the large revenues associated with extraction often requires high levels of violence and repression in the extractive enclaves as multinational companies and governments seek to guarantee the supply of natural resources though the opening up of remote frontiers and networks of connectivity.[16]

Latin America's move away from the Washington Consensus model, with its focus on finance and neoliberal governance, towards the Commodity Consensus, focused not on the re-design of the state but on enabling the large-scale export of primary products, has marked the beginning of a new political-economy order that challenges existing state and social structures and curtails democracy in the region.[17] Veltmeyer and Petras describe this turn towards natural resource extraction, which relates to a 'predatory and backward form of capitalism dominant in the nineteenth century' – the era of conquest and extractive colonialisation – as 'extractivist imperialism' or 'imperialism of the twenty-first century'.[18] Driven by the high profits associated with natural resource extraction, governments have refocused their attention on the large-scale extraction of natural materials. As Svampa remarks, 'in terms of the logic of accumulation, the new Commodities Consensus adds to the dynamic of dispossession of land, resources and territories whilst simultaneously creating new forms of dependency and domination'.[19] The commitment by Latin American governments to expand the extractive economy has led to the repoliticisation of minerals and a general unwillingness by leaders to consider demands for environmental justice or to allow civil society to play an increased role in mineral politics.[20] In his now infamous 2007 manifesto on a modern extractive economy, the former Peruvian President Alan García argued that civil society groups and environmental activists opposed to mining and natural resource exploitation were standing in the way of the country's progress and compared them to Aesop's dog-in-the-manger.[21]

Svampa argues that a combination of three axes – euphemistically described as 'sustainable development', 'corporate social responsibility' and so-called 'good' governance, have created a shared framework of the neoliberal discourse that aims to legitimise extractive economic projects. First, the concept of sustainable development has become associated with a diluted idea of sustainability, whereby the limits to growth proposed by environmentalists have been shifted. 'Development' rather 'sustainable' is the operative word in this formulation. This version of sustainability promotes an eco-efficient position that continues to view nature as capital while embracing the so-called 'green' growth and new technological 'fixes' to overcome social and environmental problems. Second, the concept of corporate social responsibility (CSR) has been widely promoted by the UN to promote voluntary compliance with human rights, environmental and labour standards and adopted by many large transnational corporations to neutralise criticism of the harmful impacts of their economic activities. CSR recognises that corporations are the primary subjects of globalised economies and they ought to be accountable for dealing with any conflicts relating to the social, economic and ecological impacts that arise as a consequence of their activities.[22] In 2011, the UN Human Rights Council adopted the Guiding Principles for Business and Human Rights. The first corporate human rights responsibility initiative to be endorsed by the UN, the framework set out three guiding principles for preventing and addressing the risk of adverse human rights impacts linked to business activity. These Guiding Principles included: states' existing obligations to respect, protect and fulfill human rights and fundamental freedoms; the role of business enterprises as specialised organs of society performing specialised functions, required to comply with all applicable laws and to respect human rights; and the need for rights and obligations to be matched to appropriate and effective remedies when breached.[23] Third, CSR has become connected to the concept of governance as a micro-political conflict resolution mechanism between multiple actors, promoting the idea that a symmetrical relationship exists between those involved and viewing the different levels of the state as another participant.[24]

Further, although the 1986 Declaration on the Right to Development was initially viewed as progressive because of its promotion of social progress, the declaration has also provided a human-rights-based justification for exploiting the world's natural resources and damaging the environment. The right to development has proved to be problematic in that it has strengthened the position of sovereign states, particularly, 'the integrity of the independence of governments over the geographic areas that states have been able to call their own' and their natural resources and natural wealth found within their geographic boundaries.[25] This strengthening of national sovereignty has and continues to hinder the emergence of a viable solution to the environmental challenges we face today and to the universal right to health as well an environment to sustain this.

Despite the region's changing political climate evidenced by the rise of left and centre-left governments in countries such as Bolivia, Brazil, Ecuador and Venezuela, this shift was also accompanied by new post-neoliberal and post-development agendas to enable these states to secure their competitive advantage in meeting the global surge in demand for raw materials. Extractivism was thereby cemented as the cornerstone of growth-oriented development policies in Latin America. The alternative platforms – neo-extractivism – put forward by these governments have purported to transcend traditional growth-centric economic models and break imperialist dependency by offering radical alternatives to the way in which socio-economic development discourses are constructed. Unlike conventional extractivism, characterised by the limited role of the state often subordinated to the interests of transnational corporations, under the framework of neo-extractivism, the state has taken on a more interventionist and regulatory role, introducing a new socio-political dimension into the practice of extractivism. Thus, while Latin America's progressive governments have created a new type of extractivism that bears an apparently 'progressive stamp' through the regulation of the appropriation of resources, increase in export duties and taxes, renegotiation of contracts and redirection of surplus revenue to social programmes, natural resource extraction has intensified.[26] However, Bebbington and Bebbington comment that, 'the troubling face of this policy convergence has been the predisposition toward authoritarian imposition of the model combining occasional use of force with efforts to delegitimise those who question extraction'.[27] This policy disposition has led to a plethora of social conflicts that are not just manifestations of struggles over human rights, forced displacement, citizenship and control over political economic processes and natural resources but, as Blaser argues, are also in defence of the 'complex webs of relations between humans and nonhumans' that for indigenous peoples are 'better expressed in the language of kinship than in the language of property'.[28] Furthermore, as Veltmeyer and Petras argue, the social and political struggles surrounding extractivism have given rise to a new class struggle predominately in rural areas. This has created a new proletariat composed of waged workers and miners, indigenous communities, peasant farmer communities and semi-proletarianised rural landless workers who form the backbone of the forces of resistance against the 'workings of capitalism and imperialism in the economic interests of the dominant class'.[29] In the face of natural resource exploitation, new theoretical, political and economic conceptualisations of the relationship between humans and the natural environment are being formulated, assessed and challenged in Latin America.[30]

Although the Washington Consensus is being questioned in Latin America, the neoliberal discourse is still very much hegemonic. Indeed, neoliberal ideology has been interwoven with neo-extractivism, the new so-called progressive development rhetoric.[31] Neo-extractivism bares the usual stamp of prioritising economic growth and national development agendas over human and environmental rights. Neo-extractivism has become 'a part of South America's own contemporary version of development, which maintains the myth of progress under a new hybridisation of culture and politics'.[32] Yet, even under its contemporary guise, neo-extractivism fails to substantially change the current structure of accumulation and move away from a productivist appropriation of nature and extractivist policies remain hegemonic in the region while the lingering and persistent problems associated with previous imperialist policies prevail. Veltmeyer and Petras argue that in opting for the resource development strategy progressive governments have done little more than strike a better deal with 'the agents of global extractive capital in a coincidence of economic interests: to share the spoils (windfall profits and enhanced claims on ground rent)'.[33] Consequently, the capitalist state remains 'at the centre of the system in its active support of extractive capital – in paving the way for the operations of extractive capital and backing up these operations with the power at its disposal'.[34] Although it was hoped that the rise of progressive governments in Latin America would lead to a transition away from extractivist activities towards a more sustainable type of development, these governments have in fact continued to maintain classic extractivism, albeit with a progressive twist. They have replaced the old extractivist discourse that pointed towards exports or the world market with one that points to globalisation and competition.[35] Moreover, the current focus on developing large-scale, export-oriented extractive projects in the region has enabled both 'progressive governments that question the neoliberal consensus and other governments that continue a conservative political agenda within the neoliberal framework' to co-exist.[36]

Ecuador and Bolivia in particular have been drawn into debates over extractivism because of their adoption of the *Buen Vivir* [live well] concept and unique inclusion and acknowledgement of the rights of nature and mother earth in their constitutions of 2008 and 2009 respectively. The uneasy marriage of extractivism and *Buen Vivir* makes neo-extractivism a particularly contradictory and complex phenomenon. A cognitive and epistemic shift has been advocated over the last two decades or so in order to move away from modernist paradigms and to adopt original epistemological and ontological narratives in which rearticulating the natural environment's role must be paramount.[37] The post-development concept of *Buen Vivir* moves beyond traditional Western development theory, based on a narrow set of indicators, transforming the relationship between development policy and social well-being. The theory and practice of *Buen Vivir* presupposes a new set of rights based on plurality and coexistence rather than on dialectical dualities and hierarchies. The *Buen Vivir* paradigm has become an integral part of Latin America's post-neoliberal policy framework and socio-economic transition, driven and articulated by the region's leftist governments and indigenous social movements.[38]

In Bolivia and Ecuador the concept has gained broad social, cultural and political support. Both states have redefined themselves as plurinational states in a post-colonial context, incorporating *Buen Vivir* principles into their national development plans and new constitutions.[39] In Bolivia, *Buen Vivir* represents the state's basic principles and orientation, promoting a pluralistic society's ethical and moral principles. It refers to the Aymara concept of Suma Qamaña and to the Guaraní ideas of *ñandereko* [harmonious living], *teko kavi* [the good life], *ivi maraei* [the land without evil] and *qhapaj ñan* [the path to a noble life], emphasising in particular the protection of *Pachamama* [Mother Earth]. The Ecuadorian conceptual framework for *Buen Vivir* differs in that it refers to plural sets of rights based on the indigenous Quechua notion of *sumak kawsay*, which includes the rights to freedom, participation, health, shelter, education, food, as well as the rights of nature, rather than an ethical principle for the state as in the case of Bolivia.[40] Furthermore, both countries have adopted the UN Declaration on the Rights of Indigenous Peoples (UNDRIP) and ratified ILO Convention 169, establishing that indigenous peoples have, among others, the right to free, prior and informed consent, the right to self-determination and self-determined development. These new constitutions and conventions have become a crucial weapon in the struggle against the old elites and also against the very same neo-extractivist governments who supported constitutional reform.[41]

While the critique of the growth-based development model has extended so far as to entail a deeper and more comprehensive critique of euro-modernity and modern ontology through *Buen Vivir's* relational ontology, what has proved to be more challenging is the realisation of these ethical and moral principles and plural sets of rights in state practice. First, as in the cases of both Ecuador and Bolivia, the contradictions inherent in the attempts to turn constitutional principles into policy are apparent first and foremost in the abundance of modernist linguistic concepts such as 'growth', 'productivity', 'efficiency' and 'market economy'. In other words, without a clear political project that implements it through effective policies, this new decolonial episteme may remain vague and often problematic.[42] Second, particularly with regard to the expansion of the hydrocarbon and mining industry as well as the construction of mega-projects and infrastructure in these countries, clear contradictions exist between the discourse of *Buen Vivir* and the current development agenda. Not only does extractivism violate *Buen Vivir's* rhetoric of harmonious living between the human and non-human, the rights of nature with regard to existence, maintenance and regeneration of its life cycles, and important equilibriums such as quality of life, democratisation of the state and a focus on biocentric concerns, but it also perpetuates the exclusion–inclusion dichotomy and hierarchical articulations, a logic that has traditionally been associated with hegemonic modernist development paradigms. Third, the deepening and extension of the extractivist development model is hampering the potential articulation of an alternative to extractivism in the region and the application of *Buen Vivir*. Addressing the political economy and changing the productive matrix is the most urgent challenge facing *Buen Vivir* today. A post-extractivist strategy is imperative to halt the acute social and environmental impacts of extractivism, address the high propensity for conflict that surrounds extractivist activities, prevent resource depletion and deal with the global ecological crisis and climate change by using nature in a rational and sustainable manner.[43]

Although the policy of extractivism under the new left governments of Latin America might be viewed as more progressive than previous forms of extraction, say in relation to the distribution of economic benefits, they still continue to support capitalist production modes through hydrocarbon expansion.[44] Gudynas argues that the persistence of conventional development is symptomatic of 'how deeply rooted and resistant to change the ideologies of "modernity" and "progress" are in our culture'.[45] Conventional extractivism and progressive neo-extractivism share key aspects in common such as 'the appropriation of nature to feed economic growth, and the idea of development understood as an ongoing, linear process of material progress'.[46] Therefore, any alternative to development must deal with extractivism and promote a post-extractivist agenda that will break and overcome dependency, an idea that has been dismissed by critics as impossible or naive. This does not suggest a ban on all extractive industries but rather a massive decrease whereby the only industries left operating are those that are essential, directly linked to national and regional economic chains, and meet social and environmental conditions. To reach this stage, economies must transition immediately from 'predatory extractivism' to 'sensible extractivism', where industries fully comply with social and environmental laws and are rigorously controlled, and finally to 'indispensable extractions' where only essential industries remain.[47] Furthermore, if the transition to post-extractivism is considered within the *Buen Vivir* framework, the process of change must meet two critical conditions: (1) poverty eradication and; (2) prevention of new losses of biodiversity. This would involve considering both environmental limits and quality of life when considering the use natural resources in the production matrix and reducing over-consumption, which contributes to poverty levels and environmental problems.[48] These two critical conditions are also central to the aspects of the post-2015 UN sustainable development agenda, though it too assumes economic growth is a must.[49]

## Human rights, rights of nature and the environment

As the global boom in commodities prices led Latin American governments to pursue extractive industry growth policies, social movements in the region have increasingly become engaged in the debate on biodiversity conservation and appropriation, as well as in redefining cultural and ethnic identities.[50] The decentring of euro-modernist perspectives has contributed to strengthening ethnic politics in the region in relation to ecology and environmentalism, and opened up critical new political spaces allowing for the expression of indigenous knowledge, traditions and cultural identity which had previously been oppressed,[51] laying the foundations of today's social and environmental struggles. Social movements in the region are increasingly questioning the epistemological frameworks based on a dialectic system of inclusion-exclusion upon which the developmentalist socio-economic model is based. The struggle over the expansion of the extractivist and neo-extractivist development models, the absence of participatory democracy, and the criminalisation of resistance have led to the rupture between the state and social movements. In countries such as Ecuador and Bolivia with the construction of the political agency of indigeneity, the state and indigenous movements have reached a critical impasse. Environmental discourses are intrinsic to indigenous cosmologies so any new political and social ecology based on alternative cosmologies rejects the modern one.

Environmental protection remains one of the most challenging issues of international law in the twenty-first century.[52] However, as a consequence of the broadening of economic and social rights to incorporate elements of environmental protection, evident in rights treaties such as the International Covenant on Civil and Political Rights (ICCPR), the International Covenant on Economic and Social and Cultural Rights (ICESCR), and the United Nations Declaration on the Rights of Indigenous Peoples (UNDRIP), the idea that people are entitled to the right to a decent environment has gained traction over recent years.[53] While it is increasingly recognised within international law that environmental degradation can deprive human rights, such as the right to health, which is determinant of a wide range of factors associated with the environment, and that 'mere recognition of such deprivations is not enough to promote and secure a healthy environment',[54] a non-derivative human right to the environment has yet to be recognised.[55] Furthermore, new visions for the relationship between human rights and the environment have not been explored within the UN system.[56] Instead, in the last two decades, human rights law has undergone a rapid greening, whereby the focus has been on reinterpreting universally recognised rights. There has been a convergence between human rights and environmental protection whereby environmental integrity is being recast as a mechanism of enforcement of human rights, 'functioning as *sine qua non* conditions of existence for the realisation of much of the human rights agenda'.[57] Consequently, three theoretical approaches to the relationship between human rights and the environment have emerged. First, the environment is seen as a precondition to the enjoyment of human rights. Second, human rights can be used as a tool to address environmental issues from both a procedural and substantive stance and lastly, human rights and the environment have increasingly been grouped together as the conditions for sustainable development.[58]

The enjoyment of economic and social human rights, such as the right to life, water, health, personal security, an adequate standard of living, tenure and resource rights and self-determination, and environmental protection are explicitly and implicitly interrelated. However, equally important in the environmental context are civil, political and procedural rights that promote access to courts and the justice system, the ability to protest and the capacity to obtain information. Gearty argues that given their central role in human rights law, civil and political rights tend to be more open to legal avenues than others. This has allowed activists 'to smuggle their true goals into law cases camouflaged as traditional legal actions concerned only with civil rights'.[59] Although the linkage between the environment and human rights was recognised internationally at the Stockholm Declaration of the United Nations Conference on the Human Environment in 1972 and later at the Rio Declaration on Environment and Development in 1992, the current global ecological crisis has led to a resurgence in interest in the connection between human rights and the environment. Most recently, human rights have featured prominently in the 2030 Agenda for Sustainable Development, with human rights, if not explicitly then implicitly, at the core of the 17 sustainable development goals and 169 targets. Moreover, the 2015 Paris Agreement on Climate Change marked a watershed moment with the pre-amble to the agreement referencing human rights and states' obligations to respect, promote and consider their respective obligations on human rights when taking action to address climate change, marking the first such reference in a multilateral environmental agreement. The UN Special Rapporteur on Human Rights and the Environment, John Knox, underlined the importance of human rights at the Paris Climate Conference, remind-ing parties that 'States' human rights obligations also encompass climate change', urging them to adopt a rights perspective in tackling environmental issues.[60]

Although, the connection between human rights and the environment has been recog-nised for some years, it is only recently that activities that harm the environment, human and non-human life, as well as the planet itself, have been thought of 'as activities that might be considered criminal or at least seriously harmful with intergenerational conse-quences and transnational impacts'.[61] Examples of environmental harm and crime have been organised into two categories and classified as either resulting directly from the destruction and degradation of the earth's resources (primary) or as being symbiotic with or dependent upon such destruction, and efforts made to regulate or prevent it (second-ary).[62] The development of a green perspective in criminology has played a critical role in rethinking human legal systems and developing alternative 'benchmarks' to legal defi-nitions of crime, including, human rights abuses and social harm as advocated by Potter, and Raftopoulos and Short.[63] As Cullinan observed, 'a primary cause of environmental destruction is the fact that current legal systems are designed to perpetuate human domina-tion of nature instead of fostering mutually beneficial relationships between human and other members of the earth community'.[64] This has resulted in the plundering of the Earth's resources though activities such as extractivism and environmental degradation and destruction. Current legal systems based on the belief that 'humans are separate from and superior to all other members of the community, and that the primary role of Earth is to serve as "natural resources" for humans to consume'[65] are failing to protect the non-human world and are perpetuating an exploitative relationship by defining the Earth's natural resources as property.[66] This has promoted calls to develop a new jurispru-dence for the Earth, whereby legal systems 'take an evolutionary leap forward by recognis-ing legally enforceable rights for nature and other-than-human beings'[67] and a proposal to the UN Law Commission in 2010 for an international law of Ecocide that would recognise human-caused environmental damage and degradation as a crime against international peace.[68]

The failure of current legal systems and environmental laws to protect both humans and the non-human world has led to one of the most important developments in the environ-mental rights revolution; the questioning of Western liberal approaches to human rights and the incorporation of intercultural perspectives which expands the notion of human dignity. Santos argues that this pragmatic transition in human rights is occurring because 'our time is witnessing the final crisis of the hegemony of the socio-cultural paradigm of western modernity', spread throughout the world through colonialism and imperialism[69] and epistemicide.[70] Hegemonic political thinking has reduced 'the understanding of the world to the western understanding of the world, thus ignoring or trivialising decisive cul-tural and political experiences and initiatives in the countries in the global South'.[71] Hence, conventional human rights conceptions have historically lacked the theoretical and analyti-cal tools to be compatible with and useful to movements of resistance that reflect alternative ideologies and contradict the liberal idea of the universality of human rights or that question the notion that human nature is individualistic, self-sustaining and fundamentally different from non-human nature.[72] As Gionolla comments, 'in terms of their relationship to the environment, mainstream human rights approaches construct the protection of the environment as being an implication of the protection of human beings'[73]. Indigenous movements in Latin America have played a critical role in moving environmental protection up the human rights agenda. Furthermore, they have led the transition towards a new approach to human rights built upon alternative cosmologies that offer an alternative conception of human dignity to the Western notion, whereby nature has inalienable rights and the false dichotomy of humans being separate and superior to the non-human world is rejected. Recognition of 'rights of nature' in countries like Bolivia and Ecuador represents a transition away from euro-modernist human rights discourses and is reflective of the 'epistemic turn' that has occurred in both the methodology and practice of critical thought in Latin America since the late 1990s. This epistemic turn questions both the historical as well as the theoretical legacy of modernist categories and led to the adoption of original epistemological and ontological narratives. Yet despite incorporating the notion of living in harmony with nature into their national constitutions and granting nature inalienable rights, Ecuador and Bolivia, are still struggling to overcome the legacy of modernist development paradigms continually reinforced by the state.

The Commodity Consensus has led to a new cycle of protests that look to transcend traditional ideological and class divisions and unite around the negative impact of extractive industries, notions of development, territorial sovereignty and the defence of the commons and biodiversity. However, as the articles in this special edition demonstrate, with the state as a guardian of both human and environmental rights, these rights remain fragile due to states' continued and active support of extractivist activities. Although it has long been recognised that it is states' responsibility to protect and promote human rights in their territories, they also have a duty to adopt an appropriate and effective regulatory framework in order to prevent human rights abuses. While human rights may have been strengthened on an international level, they can only be realised if individual states guarantee and enforce human rights agreements domestically. States have an obligation to monitor and supervise extraction, exploitation and development activities, guarantee mechanisms of effective participation and access to information, prevent illegal activities and forms of violence, as well as guarantee access to justice through investigation, punishment and adequate reparations for violations of human rights committed under these circumstances.[74] Yet, because of Latin American governments' commitment to pursuing and intensifying extractivism, the role of the state is focused not on acting as a guarantor of human rights but rather on protecting and facilitating their own economic interests. As Veltmeyer and Petras remark:

> Because of the coincidence of economic interests between the state and capital (resource rents for the governments, profits for the companies), governments in the region – even those oriented towards a policy of anti-imperialist capital, and in any conflict between the company and the communities directly affected by the operations of extractive capital these governments tend to side with capital against the communities.[75]

The self-serving nature of the state has severely impacted on the ability of human rights discourses to curb and combat environmental degradation in Latin America. Governments throughout the continent have taken steps to limit the effects of social mobilisation against extractivism by introducing measures to limit their economic, social, political and civil rights such as to participate in decision-making, acquire information, freedom of expression, and freedom of assembly. A new report by ARTICLE 19, CIEL, and

Vermont Law School, criticised Latin American states for the increasing criminalisation of protests, and the use of the law, such as anti-terrorism legislation and libel threats, to quell dissent against extractivist activities.[76] In Peru, authorities have increased penalties for committing a public order offence, made it easier for the military to intervene in social-environmental conflicts, and supported impunity for official abuses. Ecuadorian authorities have limited the right to freedom of assembly by requiring protest organisers to gain per-mission from the municipality and police superintendent to hold a protest and criminialising through imprisonment or fines demonstration leaders without the relevant paperwork. Moreover, in Bolivia, government officials now have the power to dissolve any nongovern-mental organisations without using any judicial process.[77]

Despite the adoption of the UN resolution requiring states to ensure the rights and safety of human rights defenders, it has become increasingly clear that states are not doing enough to protect those lives at risk from harassment and violence and to bring those responsible to justice. In many cases the state or security forces are perpetuating violence against those groups opposed to natural resource extraction. Moreover, little is being done to protect mar-ginalised and vulnerable communities whose livelihoods, cultures and identities are suscep-tible to environmental and social consequences of extractivism and also to protect the environment from further harm and degradation. As the contributions indicate, environment protection, the rights of nature and human rights are continuously sidelined in the name of economic development, even in those countries with a more progressive development agenda. However, while the application of human rights discourses has yielded limited con-crete results, human rights have provided a 'language of protest' and a 'platform for change'[78] for those communities and social movements struggling against the expansion of extractivist activities in Latin America. The use of human rights has become an important means of exposing both the ecological and social destruction that accompanies many extra-ctivist projects and has ultimately broadened the frame of both action and discourse sur-rounding socio-environmental conflicts while simultaneously increasing the attention focused on human rights and the rights of nature. As the recent groundbreaking case of the Embera Chamí people of the indigenous Resguardo Cañamomo Lomaprieta in western Colombia demonstrates, human rights discourses offer hope and a reason to remain optimistic. In February 2017, the Colombian Constitutional Court granted the peti-tion for the protection of constitutional rights requested by the Embera Chamí people and ordered their lands to be delimited and titled within one year, during which time all further permits or formalisation of mining activities must be suspended. Furthermore, any sub-sequent mining activities may only proceed with the full cooperation and consent of the Resguardo.[79]

## Social-environmental conflicts in Latin America

This special issue of the *International Journal of Human Rights* focuses on the issues of global environmental injustice and human rights violations and explores the scope and limits of the potential of human rights to influence environmental justice. It offers a multi-disciplinary perspective on contemporary development discussions, analysing some of the crucial challenges, contradictions and promises within current environmental and human rights practices in Latin America. Taking a multi-level perspective that links the local, national, regional and transnational levels of inquiry, each contribution approaches ques-tions concerned with human rights and environmental justice from a variety of theoretical and methodological viewpoints. The contributors examine how the extraction and exploita-tion of natural resources and the further commodification of nature have affected local communities in the region and how these policies have impacted on the promotion and protection of human rights as communities struggle to defend their rights and territories. Bringing together scholars from diverse disciplines such as sociology, political science, anthropology and social science, this special issue analyses the emergence of transnational activism in the context of collective action organised around socio-environmental conflicts, the infringement of basic human rights and the emergence of alternative and sometimes conflicting development models. Furthermore, it critically discusses why governments are often willing to override their commitments to sustainability and human rights to promote their development agenda.

Joanna Morley in her article "' ... Beggars sitting on a sack of gold": Oil exploration in the Ecuadorian Amazon as *buen vivir* and sustainable development' analyses the tensions within *Buen Vivir*, an innovative interpretation of the concept of sustainable development, by examining the practice of human rights in socio-environmental conflicts in Ecuador. Morley argues that the contradictions that exist between the rhetoric of *Buen Vivir* and President Correa's neo-extractivist development agenda mirror those that exist within the rhetoric of social inclusion, environmental protection and sustainable economic growth found in Agenda 2030. Moreover, the expansion of extractivist activities in Ecuador reflects the pragmatic arguments that to be effective and politically acceptable, development and environmental approaches must develop strategies that work with the economic interest mechanism of the neoliberal framework of industrialised countries. Morely questions whether a sustainable development agenda that seeks to decouple economic growth from development and in which economic, social and environment development are viewed as equal is feasible in the current neoliberal model of global governance. Examining oil exploration in the southern Ecuadorian Amazon, she discusses how the Ecuadorian state, despite having passed a progressive constitution and promoting a development model that has the potential to be a realistic alternative to neoliberal capitalism, is willing to override and marginalise the rights of their traditional supporters in order to pursue economic growth.

Radosław Powęska in his article 'State-led extractivism and the frustration of indigenous self-determined development: lessons from Bolivia', discusses the incorporation of indigenous rights and the problems associated with their genuine implementation in Bolivia in the context of state-led extractivism. Powęska questions to what extent those human rights can be an effective tool against extractive enterprises harmful to the interests of indigenous peoples, as well as the very relationship between extractivism and the employment of human rights in Bolivia. He analyses the role of the character of the state and other related internal factors impacting on the viability of indigenous rights related to self-determination and development, focusing in particular on the political culture and historically developed state–society relations, based upon and reflecting the asymmetries of power and inequalities. Powęska begins discussing the paradox of the rhetoric of human rights, in that although indigenous rights are being strengthened through international activism on the global level, their implementation strictly depends on local circumstances. Questioning the authenticity of its pro-indigenous agenda, he argues that the Bolivian state fails to protect indigenous rights despite its promises and indeed promotes extractivism instead, because of the central role of resource exploitation in generation of rents that fuel paternalist–clientelist state–society relations and help to reproduce power structures. Furthermore, the imposition of extractivist-based development on indigenous communities is a negation of their right to self-determination and indigenous rights and the indigenous agenda in Bolivia is being deformed and manipulated by the state.

Taking a political economy perspective on the extractive dilemma in Bolivia, Rickard Lalander in his article 'Ethnic rights and the dilemma of extractive development in plurinational Bolivia' examines the tensions between ethically defined rights in relation to broader human rights in terms of values and norms related to welfare. The article contributes to debates on contentious resource governance and the relationship, contradictions and tensions between class and ethnicity amid Bolivian identity politics and the question of indigeneity. Lalander argues that despite the Bolivian Constitution of 2009 being one of the most radical in the world with regard to the incorporation and recognition of human rights and indigenous rights, in practice, class-based human rights tend to be superior to those ethically defined because of the extractive development dilemma. He examines both the complex identity politics of Bolivian indigeneity and the extractive dilemma of Evo Morales' government, in particular, the discourse and moral justification for the implementation of extractive politics and how these discourses and justifications relate to the identitarian elements of class and/or identity. Using the TIPNIS conflict as a case study, Lalander illustrates the contradictions that exist between indigenous rights claims and state practices, such as those indigenous rights reinforced in the 2009 Constitution which continually clash with the rights of the nation state to extract and commercialise natural resources.

Marieke Riethof in her article 'The international human rights discourse as a strategic focus in socio-environmental conflicts: the case of hydro-electric dams in Brazil' discusses how human rights discourses have become a powerful moral and political resource to critique the social impact of Brazil's development agenda. Riethof examines the mobilisation of human rights campaigns against hydro-electric dams and argues that the symbolic and legal power of human rights has allowed activists to challenge official accounts of the impact of dams on communities and the environment while deploying domestic and international legal frameworks. Anti-dam mobilisations have used human rights as a platform to highlight the discrepancies between Brazil's ambitions for global leadership within the arenas of environmental sustainability and human rights, and the domestic realities. However, although dam construction sites in Brazil have become significant sites of contestation and activists have channelled human rights discourses, the politicisation of natural resources has severely limited the space for opposition to be heard and the effectiveness of anti-dam mobilisations. Riethof concludes that while the power of the national development discourse in Brazil has restricted the debate on procedural and substantive issues, the employment of international human rights discourses and legal strategies, while unable to halt dam construction, have been able to exert political pressure on the government in a polarised context.

John-Andrew McNeish, in his article 'Extracting justice? Colombia's commitment to mining and energy as a foundation for peace', considers the idea that natural resource extraction can pay for peace and justice. Extraction has been advanced as a vital source of funding to cover the costs of an eventual peace and continued economic development, so the Colombian state has taken a number of steps in recent years to ease and simplify the environmental licences in order to further develop the mining and energy sector and crack down on illegal extractive installations. In doing so, the Colombian state has deliberately flouted constitutionally founded principles that support popular sovereignty and local democratic governance of the environment. Efforts to halt the expansion of natural resource extraction have been meet by state-led violence and the abuse of human rights. McNeish argues that the idea that the extractive sector will represent a route to justice is extremely flawed and the insecurities caused by the extractive economy will continue into the post-conflict period. Moreover, the expansion of natural resource extraction has fuelled the mutation of the armed conflict, which is now as much about oil and minerals as it is about land, political ideology and coca production, resulting in more economic uncertainty, the largest internally displaced population in the world and the mass abuse of human rights. Therefore, even with the signing of the Peace Accords, the current legal and socio-economic dynamic indicate that much of the violence linked to natural resource extraction will continue into the foreseeable future, increasing insecurities and initiating a new phase of human rights violations.

## Extractivism and human rights: new engagements

It has become increasingly apparent that the Commodity Consensus model and the large-scale export of primary products in Latin America have advanced in recent years in a context of increasing violence and have impacted enormously on the promotion and protection of human rights. As a consequence of this new cycle of protests in the region, the environment has emerged as a new political battleground for human rights, and along with it, the urgent need to carry out more research on the relationship between human rights, extractivism and the environment. As Bebbington acknowledged, the academic world was caught by surprise by the speed and scale of the commodity boom in Latin America, leaving academics to play 'catch-up'.[80] However, in recent years, research into extractivism has gathered pace. As demonstrated in this opening contribution, the explosion of social-environmental conflicts that have accompanied the growth and diversification of extractivist activities has posed a challenge to the political and economic ontology of current development models and opened up debates about nature and the relationship between the human and non-human world. Moreover, it has raised questions over Western approaches to human rights and led the transition towards de-colonial approaches to human rights built upon alternative cosmologies and intercultural perspectives, whereby nature has inalienable rights. Consequently, there are a number of emerging themes that warrant further attention. Further research into how transnational human and environmental rights advocacy networks are shaping the meaning and possibility of human rights discourses, de-colonial approaches to human rights and methodologies in Latin America, the adoption of human rights discourses in different social and cultural contexts and legal systems and also gendered impacts of extractivism and the role of women in social-environmental conflicts could provide valuable new insights into the merits of extractivism as a development strategy. It is hoped that this special edition will not only synthesise current work on human rights and extractivism in Latin America but also encourage more multidisciplinary research into the topic, broadening the analytical base of debates on extractivism, help foster a new relationship between humans and nature and change the way we conceive the environment.

## Disclosure statement

No potential conflict of interest was reported by the author.

## ORCID

*Malayna Raftopoulos* ⓘ http://orcid.org/0000-0002-5619-8496

## Notes

1. Henry Veltmeyer and James Petras, *The New Extraction: A Post-Neoliberal Development Model or Imperialism of the Twenty-First Century* (London and New York: Zed Books, 2014), 1.
2. Phillipe Descola, *The Ecology of Others* (Chicago, IL: The University of Chicago Press, 2013), 81.
3. Inter-America Commission for Human Rights, *Indigenous Peoples, Afro-Descendent Communities, and Natural Resources: Human Rights Protection in the Context of Extraction, Exploitation, and Development Activities* (2015), http://www.oas.org/en/iachr/reports/pdfs/ExtractiveIndustries2016.pdf.
4. United Nations Human Rights Council, *Report of the Special Rapporteur on the Rights to Freedom of Peaceful Assembly and of Association*, A/HRC/29/25 (2015), 15 January 2017, http://www.ohchr.org/EN/Issues/AssemblyAssociation/Pages/AnnualReports.aspx.
5. John-Andrew McNeish, 'More than Beads and Feathers: Resource Extractions and the Indigenous Challenge in Latin America', in *New Political Spaces in Latin American Natural Resource Governance*, ed. Håvard Haarstad (Basingstoke: Palgrave Macmillan, 2012), 39–60, 41.
6. Global Witness, *On Dangerous Ground* (June 2016), 1–27, 3, https://www.globalwitness.org/en/reports/dangerous-ground/.
7. United Nations Human Rights Council Resolution, 'Promotion and Protection of all Human rights, Civil, Political, Economic, Social and Cultural Rights, Including the Right to Development', A/HRC/31/L.7/Rev.1 (2016), http://www.un.org/ga/search/view_doc.asp?symbol=A/HRC/31/L.7/Rev.1.
8. United Nations Human Rights Office of the High Commissioner, *A Deadly Undertaking – UN Experts Urge All Governments to Protect Environmental Rights Defenders*, http://www.ohchr.org/EN/NewsEvents/Pages/DisplayNews.aspx?NewsID=20052#sthash.3hNjeW3t.dpuf.
9. Ana Grear, 'The Vulnerable Living Order: Human Rights and the Environment in a Critical and Philosophical Perspective', *Journal of Human Rights and the Environment* 2, no. 1 (2011): 23–44, 23.
10. Susana Borrás, 'New Transitions from Human Rights to the Environment to the Rights of Nature', *Transnational Environmental Law* 5, no. 1 (2016): 113–43.
11. Gearty, 'Do Human Rights Help or Hinder Environmental Protection?', 7.
12. Håvard Haarstad, 'Extracting Justice? Critical Themes and Challenges in Latin American Natural Resource Governance', in *New Political Spaces in Latin American Natural Resource Governance*, ed. Håvard Haarstad (Basingstoke: Palgrave Macmillan, 2012), 1–16, 1.
13. Marianne Schmink and José Ramón Jouve-Martín, 'Contemporary Debates on Ecology, Society and Culture in Latin America', *Latin America Research Review* 46 (Special Edition, 2011): 3–10, 3.
14. Alberto Acosta, *La Maldición de la Abundancia* (Quito: Abya-Yala, 2009).
15. Ibid., 136.
16. Eduardo Gudynas, 'The New Extractivism of the 21st Century: Ten Urgent Theses about Extractivism in Relation to Current South American Progressivism', *Americas Program Report* (Washington, DC: Center for International Policy, 2010), 1–14.
17. Maristella Svampa, 'Resource Extractivism and Alternatives: Latin American Perspectives on Development', in *Beyond Development: Alternatives Visions from Latin America*, ed. Miriam Lang and Dunia Mokrani (Quito: Rosa Luxemburg Foundation, 2013), 117–45.
18. Veltmeyer and Petras, 'The New Extraction', 21.
19. Svampa, 'Resource Extractivism and Alternatives', 118.
20. Barbara Hogenboom, 'Depoliticized and Repoliticized Minerals in Latin America', *Journal of Developing Studies* 28, no. 2 (2012): 133–58.
21. Alan García, *El Sindrome del Perro del Hortelano* (2007), http://www.justiciaviva.org.pe/userfiles/26539211-Alan-Garcia-Perez-y-el-perro-del-hortelano.pdf.
22. Svampa, 'Resource Extractivism and Alternatives', 122–3.
23. United Nations Human Rights Office of the High Commissioner, 'Guiding Principles on Business and Human Rights' (2011),http://www.ohchr.org/Documents/Publications/GuidingPrinciplesBusinessHR_EN.pdf.

24. Svampa, 'Resource Extractivism and Alternatives', 122–3.
25. Conor Gearty, 'Do Human Rights Help or Hinder Environmental Protection?', *Journal of Human Rights and the Environment* 1, no. 1 (2010): 7–22, 9.
26. Gudynas, 'The New Extractivism of the 21st Century', 3.
27. Denise Humphreys Bebbington and Anthony Bebbington, 'Post-What? Extractive Industries, Narratives of Development and Socio-Environmental Disputes Across the (Ostensibly Changing Andean Region)', in *New Political Spaces in Latin American Natural Resource Governance*, ed. Håvard Haarstad (Basingstoke: Palgrave Macmillan, 2012), 17–37, 19–20.
28. Mario Blaser, 'Notes Towards a Political Ontology of "Environmental Conflicts"', in *Contested Ecologies: Dialogues in the South on Nature and Knowledge*, ed. Lesley Green (Cape Town: HSRC Press, 2013), 13–27.
29. Veltmeyer and Petras, 'The New Extraction', 46.
30. Michela Coletta and Malayna Raftopoulos, 'Whose Natures? Whose Knowledges? An Introduction to Epistemic Politics and Eco-Ontologies in Latin America', in *Provincialising Nature: Multidisciplinary Approaches to the Politics of the Environment in Latin America*, ed. Michela Coletta and Malayna Raftopoulos (London: Institute of Latin American Studies, School of Advanced Studies, University of London, 2016), 1–17.
31. Svampa, 'Resource Extractivism and Alternatives', 119.
32. Gudynas, 'The New Extractivism of the 21st Century', 13.
33. Veltmeyer and Petras, 'The New Extraction', 28.
34. Ibid., 2.
35. Gudynas, 'The New Extractivism of the 21st Century'.
36. Svampa, 'Resource Extractivism and Alternatives', 120.
37. Coletta and Raftopoulos, 'Whose Natures? Whose Knowledges?'
38. Tara Ruttenburg, 'Wellbeing Economics and Buen Vivir: Development Alternatives for Inclusive Human Security', *Wellbeing Economics and Buen Vivir* 18 (2013): 68–93.
39. Thomas Fatheuer, *Buen Vivir: A Brief Introduction to Latin America's New Concepts for the Good Life and the Rights of Nature* (Berlin: Heinrich Boll Stiftung Publication Series on Ecology, 2011), 17.
40. Eduardo Gudynas, 'Buen Vivir: Today's Tomorrow', *Development* 54, no. 4 (2011): 441–7, 443.
41. Roger Burbach, Michael Fox, and Federico Fuentes, *Latin America's Turbulent Transitions: The Future of Twenty-First-Century Socialism* (London: Zed Books, 2013).
42. Arturo Escobar, 'Latin America at a Crossroads', *Cultural Studies* 24, no. 1 (2010): 1–65.
43. Eduardo Gudynas, 'Transitions to Post-Extractivism: Directions, Options, Areas of Action', in *Beyond Development Alternative Visions in Latin America*, ed. Miriam Lang and Dunia Mokrani (Quito, Ecuador: Rosa Luxemburg, 2013), 165–88.
44. Gudynas, 'The New Extractivism of the 21st Century'.
45. Gudynas, 'Transitions to Post-Extractivism', 168.
46. Ibid., 165.
47. Ibid., 175.
48. Ibid., 169.
49. United Nations Sustainable Development Goals, *Transforming Our World: The 2030 Agenda for Sustainable Development*, https://sustainabledevelopment.un.org/post2015/transformingourworld.
50. Arturo Escobar, 'Whose Knowledge, Whose Nature? Biodiversity, Conservation, and the Political Ecology of Social Movements', *Journal of Political Economy* 5 (1998): 53–82.
51. Radcliffe et al., 'Development and Culture'; Gudynas, 'Buen Vivir: Today's Tomorrow'.
52. Christiano Gianolla, 'Human Rights and Nature: Intercultural Perspectives and International Aspirations', *Journal of Human Rights and the Environment* 4, no. 1 (2013): 48–78.
53. Alan Boyle, 'Human Rights and the Environment: Where Next?', *The European Journal of International Law* 23, no. 3 (2012): 613–42.
54. Klaus Bosselmann, 'Environmental and Human Rights in Ethical Context', in *Research Handbook on Human Rights and the Environment*, ed. Ana Grear and Louis J. Kotzé (Cheltenham: Edward Elgar Publishing Limited, 2015), 531–50, 531.
55. Lynda Collins, 'The United Nations, Human Rights and the Environment', in *Research Handbook on Human Rights and the Environment*, ed. Grear and Kotzé, 219–44, 241.
56. Boyle, 'Human Rights and the Environment', 617.

57. Gearty, 'Do Human Rights Help or Hinder Environmental Protection?', 13.
58. Boyle, 'Human Rights and the Environment', 617.
59. Gearty, 'Do Human Rights Help or Hinder Environmental Protection?', 17.
60. United Nations Human Rights Office of the High Commissioner, *COP21: States' Human Rights Obligations Encompass Climate Change – UN Expert*, http://www.ohchr.org/EN/NewsEvents/Pages/DisplayNews.aspx?NewsID=16836&LangID=E#sthash.HnlZV.
61. Polly Higgins, Damien Short, and Nigel South, 'Protecting the Planet: A Proposal for a Law of Ecocide', *Crime Law Social Change* 59 (2013): 251–66, 251.
62. Ibid., 252.
63. Gary Potter, 'What is Green Criminology?' *Sociology Review* (2010): 8–12; Malayna Raftopoulos and Damien Short, 'A New Benchmark for Green Criminology: The Case for Community-Based Human Rights Impact Assessment of REDD+', in *Greening Criminology in the 21st Century: Contemporary Debates and Future Directions in the Study of Environmental Harm*, ed. Matthew Hall, Tanya Wyatt, Nigel South, Angus Nurse, Gary Potter, and Jennifer Maher (London and New York: Routledge, 2016), 165–82.
64. Cormac Cullinan, 'Earth's Jurisprudence: From Colonization to Participation', in *State of the World*, ed. Worldwatch Institute (2010), 142–8, http://blogs.worldwatch.org/transformingcultures/wp-content/uploads/2010/07/Earth-Jurisprudence-From-Colonization-to-Participation-Cullinan.pdf.
65. Ibid., 143.
66. Ibid., 144.
67. Ibid., 145.
68. Higgins et al., 'Protecting the Planet'.
69. Boaventura de Sousa Santos, 'If God Were a Human Rights Activist: Human Rights and the Challenge of Political Theologies: Is Humanity Enough? The Secular Theology of Human Rights', *Law, Social Justice & Global Development Journal* 1 (2009): 1–42, 1.
70. Boaventura de Sousa Santos, *Epistemologies of the South: Justice against Epistemicide* (Boulder, CO: Paradigm Publishers, 2014).
71. Ibid., 4.
72. Ibid.
73. Cristiano Gionolla, 'Human Rights and Nature: Intercultural Perspectives and International Aspirations', *Journal of Human Rights and the Environment* 4, no. 1 (2013): 58–78, 62.
74. Inter-America Commission for Human Rights, 'Indigenous Peoples, Afro-Descendent Communities'.
75. Veltmeyer and Petras, 'The New Extraction', 248–9.
76. ARTICLE 19, CIEL, and Vermont Law School, 'A Deadly Shade of Green' (2016), 1–72, http://www.ciel.org/wp-content/uploads/2016/08/Deadly_shade_of_green_English_Aug2016.pdf.
77. Ibid.
78. Gearty, 'Do Human Rights Help or Hinder Environmental Protection?', 7.
79. Forest Peoples Programme, 'Groundbreaking Win for Indigenous People in Colombia', 9 February 2017, http://www.forestpeoples.org/topics/mining/news/2017/02/groundbreaking-win-indigenous-people-colombia.
80. Anthony Bebbington, 'Political Ecologies of Research Extraction: Agendas Pensientes', *European Review of Latin America and Caribbean Studies*, 100 (50th Anniversary Special Issue) (2015): 85–98, 86.

# ' … Beggars sitting on a sack of gold': Oil exploration in the Ecuadorian Amazon as *buen vivir* and sustainable development

Joanna Morley

This article analyses the tensions within a concept of sustainable development, by examining the practice of human rights in the socio-environmental conflicts surrounding Ecuadorian President Correa's expansion of oil exploration in the southern Ecuadorian Amazon. Contradictions between the rhetoric of *buen vivir* and the neo-extractive development policies of President Correa mirror contradictions in the rhetoric of social inclusion, environmental protection and sustainable economic growth found in Agenda 2030, and the reality of world politics dominated by foreign capital which heads disproportionately towards extractive sectors to meet the growing energy consumption needs of industrialised countries. Correa's pursuit of further oil exploration reflects the pragmatic argument that to be effective and politically acceptable, development and environmental approaches need to develop strategies that work with the economic interest mechanisms of the neoliberal framework of industrialised countries.

While Agenda 2030 reaffirms that every state has and shall freely exercise full permanent sovereignty over all its wealth, natural resources and economic activity, sovereignty in the context of a cycle of dependence on natural resource exploitation may lead to the violation of human rights in the pursuit of economic development. Therefore, we have to question whether a sustainable development agenda that seeks to de-couple economic growth from development and in which all three pillars – economic, social and environmental development – are equal, is actually workable in the current neoliberal model of global governance?

## Introduction

To a large extent, struggles over the extraction and exploitation of natural resources define the controversial nature of contemporary Latin American politics.[1] As noted in the report of the United Nations special rapporteur on the rights to freedom of peaceful assembly and of association (FOAA), increased demand for natural resources[2] has resulted in the opening up of more areas for exploration and exploitation, especially in populated areas, leading to conflict between competing interests,[3] while citizen engagement in the natural resources sector is notoriously difficult, presenting heightened risks of human rights abuses because the sector is especially lucrative.[4]

In the era of sustainable development policies, socio-environmental conflicts involving multiple actors – including governments, local communities, indigenous populations, national or transnational companies, transnational organisations and non-governmental and academic organisations – are at the centre of Latin American development agendas and re-articulation processes more than ever before.[5] Characterised by their complexity, their varied subjects and the great diversity of the stakeholders involved,[6] conflicts surrounding natural resource extraction can be linked to transformations in the economic, institutional and ideological forces of contemporary globalisation, including the global political environment in which state-level processes are embedded.[7]

The neoliberal promotion of a 'one size fits all' approach to development policies for Latin American countries is now being tested in debates on sustainable development and climate change.[8] Latin America is 'host to an emerging consensus ... that the sustainable development agenda of tomorrow calls for a paradigm shift [and] for structural change that puts equality and environmental sustainability front and centre'.[9] The current development model, wholly dependent on the use of energy and natural resources, is environmentally degrading and will be unable to generate income growth without impairing the planet's resilience and survival.[10] In this context, Latin America has become a laboratory of initiatives seeking both ways of assessing the environmental cost of production and the financing of greener forms of growth. These initiatives include investing in renewable energy or green industries such as wave or wind power, eco-taxes, payments for environmental services, eco-tourism and biodiversity prospecting.[11]

This article aims to locate the promiscuous concept of sustainable development – 'whose claims are projected across the broadest of analytical and phenomenological boundaries' – within the practice of human rights in Latin America, where meanings are constituted by a range of social actors in the dis-articulated practices of everyday life.[12] On the one hand, ideas of the good life (*buen vivir*) and the value of Mother Earth (*Pacha Mama*) as well as granting constitutional rights to nature as in the case of Bolivia and Ecuador, represent radical innovations in the way that sustainable development can be understood. At the same time, across Latin America governments have placed great weight on generating economic growth by way of state interventions in the development of the natural resource sector, preparing the ground for market actors as a means of improving the lives and livelihoods of the majority of the population.[13]

Latin America has one of the greatest endowments of natural capital in the world[14] and natural resource-based industries are central to the economic structure of most of the region's economies.[15] The continent has 20% of the world's forested area, 15% of the planet's arable land[16] and nearly one-third of the world's total fresh water reserves.[17] In terms of subsoil resources, the continent holds 20% of world petroleum reserves, as well as 25% of the world's biofuel reserves, 44% of the world's copper, nearly 50% of silver, 65% of lithium, 33% of tin, and 22% of iron. Activities based in the exploitation of these resources are considered to be problematic for sustainable development as they produce concentration, low inclusion and environmental damage.[18]

At the same time, however, we cannot ignore that poverty in the region was cut by nearly half during the last decade,[19] more than 70 million people were lifted out of poverty, and Latin America and the Caribbean (LAC) reduced inequality[20] by five percentage points on the regional Gini index.[21] Yet recent data have shown a 'new normal' of stagnant growth rates across Latin America[22] and that inequality reduction is plateauing across the region.[23] As highlighted by the United Nations Development Programme (UNDP), in LAC the number of poor has risen for the first time in a decade. This means that three million people in the region fell into poverty between 2012 and 2014 and that economic growth is not enough to build resilience or the ability to absorb external shocks such as financial crises or natural disasters.[24]

Policymakers have warned that in the current context of a 'cycle of dependence'[25] on natural resources for economic growth and of falling exports and lower international commodity prices, Latin America is at a crossroads: 'The export model is exhausted in economic terms as the welfare model is in social terms.'[26] This has led to a common view that LAC countries should heavily tax natural resources, encourage other economic activities, and the region should induce structural change away from these industries towards more knowledge intensive sectors.[27] This includes attracting investment, investing efficiently and using criteria of social and environmental sustainability to move beyond the 'extractivist paradigm' towards productive diversification. The aim is to bring about a transformation of capital, away from the region's non-renewable resources and towards human capital, such as education and capacity-building, physical and social infrastructure, and innovation and technological development.[28]

This aim is also reflected in the concept of sustainable industrialisation and value addition prescribed in Sustainable Development Goal 9, intended to maximise the developmental impact of natural resources for inclusive and sustainable industrialisation based on increased resource-use efficiency and greater adoption of environmentally sound technologies and industrial processes.[29] Sustainable industrialisation also includes quality and resilient infrastructure to build dynamic, sustainable, innovative and people-centred economies that increase productive capacities, productivity and productive employment and financial inclusion.[30] However, the dependence of developing countries on foreign capital enables foreign direct investment (FDI), which heads disproportionately towards extractive sectors, to retain considerable influence over the terms of debate about the future direction of sustainable development policies.[31]

Making extractives the centrepiece of development strategy poses questions about long term sustainability, as resources deplete and global commodity prices fluctuate.[32] By studying the practice of human rights in the socio-environmental conflicts surrounding Ecuadorian President Correa's expansion of oil exploration in the southern Ecuadorian Amazon, the implications of the tensions within a concept of sustainable development that promotes the environmental and social pillars of development as equal to economic development can be examined.

## 1. *Buen vivir* as sustainable development

The primary policy response to poverty and environmental issues in the international arena has been the pursuit of sustainable development policies,[33] as codified in the Sustainable Development Goals (SDGs) adopted by the international community in December 2015.[34] According to United Nations Secretary General (UNSG) Ban Ki Moon

> The Sustainable Development Goals can only be reached through national ownership and local initiative ... through policy innovation and integration. It is a reminder that we will have to be creative in linking the three dimensions [social, economic and environmental] of sustainable development.[35]

For the scientific community and development practitioners the success of a development model based on the principles of *buen vivir* (living well) – a central tenet of the indigenous cosmovision in the Andes[36] – is that it proposes an alternative development paradigm in response to the failure of orthodox growth models.[37] Core concepts of *buen vivir*

(nature, community, labour, *Ayllu*, consensus and democracy, spirituality), and some fundamental principles (reciprocity, complementarity and relationality), have acquired essential
connotations that have far- reaching implications when it comes to drafting, planning
and implementing development strategies.[38]

In Ecuador, the evolving relationship between the state and oil hydrocarbon industries is
giving shape to changing economic development and territorial configurations, especially
in the oil producing areas of the Ecuadorian Amazon.[39] Increased oil exploration in the
Amazon rainforest and on traditional indigenous lands is a permanent point of disagreement
between Correa's government and indigenous social organisations.[40]

Correa's developmentalist version of *buen vivir* based on neo-extractivism departs significantly from the popular demands that brought the indigenous concept of *buen vivir* (or in
Kichwa *sumak kawsay*) into the political sphere to push for a return to use values and con-
vivial living.[41] In this context, international debates surrounding Ecuador's Yasuní-ITT
Initiative – the embodiment of *buen vivir* and the incommensurability between environmental protection, indigenous rights and environmental resources[42] – and its subsequent
cancellation are central to the post-2015 sustainable development agenda, namely how to
reduce poverty and inequality while at the same time protecting the planet and mitigating
climate change.

The government of Ecuador adopted the Yasuní-ITT Initiative (the Initiative) in April
2007 after Albert Acosta, Minister for Energy and Mines, put forward the Initiative as a
counterproposal to Petroecuador's plan to exploit concession block 43 containing the Ish-
pingo, Tiputini and Tambococha (ITT) oil fields in the Yasuní National Park.[43] The Initiative was promoted as a commitment to refrain indefinitely from extracting 846 million
barrels of oil reserves in the ITT oil fields (nearly a quarter of Ecuador's total reserves),
in exchange for international funds equivalent to half the value of the estimated oil revenues, and as an innovative option for combatting global warming, protecting the biodiversity
of Ecuador, and for supporting the voluntary isolation of the indigenous peoples living in
the Yasuní National Park.

The Initiative recognised the rights of nature to exist according to the Ecuadorian constitution[44] and the rights of humans to live within healthy environments that are increasingly threatened by the global race to drill for fossil fuels in ever more remote and
pristine areas.[45] In presenting the Initiative to the UN General Assembly (UNGA) on 24
September 2007, Correra declared

> This would be an extraordinary example of global collective action that would not only reduce
> global warming … but also introduce a new economic logic for the twenty-first century …
> [that] recognises the use and service of non-chrematistic values of environmental security
> and maintenance of world biodiversity.[46]

It moved the global conversation of sustainability out of the theoretical confines of international conferences to address the reality of indigenous peoples and their communities,
including the cultural choices of peoples in voluntary isolation.[47]

Ecuador promoted the Yasuní-ITT Initiative within the concept of 'shared but differen-
tiated responsibilities' as per the Rio Declaration on Environment and Development,[48] and
proposed that Yasuní Guarantee Certificates be recognised in the carbon market. By linking
the Initiative to the United Nations Framework Convention on Climate Change
(UNFCCC), the challenge for the Initiative was to convince leading industrialised
nations to go beyond traditional market 'off-set' investments and above-ground measures,
and to accept the concept of avoided emissions from unexploited fossil fuel reserves.[49] By

leaving oil unemitted and in the ground, and by enhancing the participation of developing countries and indigenous groups in climate change mitigation,[50] the Initiative directly confronted criticisms about the questionable commitments to environmental integrity, equity and indigenous rights within the Clean Development Mechanism[51] (CDM) and the programme for Reduction of Emissions of Deforestation and Degradation (REDD).[52]

The Yasuní-ITT fund was expected to reach at least $3.6 bn in a 13-year period with voluntary contributions from national governments, international organisations, private corporations, non-governmental organisations (NGOs), and individuals. Interest earned from the fund would be invested by the state, in line with the National Development Plan (*Plan Nacional*) and the Millennium Development Goals, to conserve one million hectares of forest and to promote social development and the transition to a new development strategy based on equality and sustainability.[53] The proposal received significant support from international institutions, European governments, international NGOs, scientific communities and personalities worldwide.[54]

In 2010, after three years of technical consultation, an international trust agreement was signed by Ecuador and the United Nations and a major international fundraising campaign was launched. At the beginning of 2012, with $116 m pledged,[55] the Ecuadorian government announced that it would move forward with the Initiative. And yet, in August 2013 President Correa signed an executive decree to cancel the Initiative, blaming the failure on the lack of foreign support (the trust fund received only $13 m in deposits).

Correa announced 'The world has failed us … it was not charity that we sought from the international community, but co-responsibility in the face of climate change.'[56] In a statement, the UNDP announced that the initiative 'was born as a national proposal, and the decision to conclude it is also a national prerogative.'[57]

The cancellation of the Initiative prompted mass demonstrations around the country, leading to the formation of the grassroots movement Yasunidos,[58] who, in April 2014, handed in a petition with approximately 850,000 signatures to attempt to trigger a referendum on drilling in the ITT block. Ten days later the National Electoral Council (NEC) announced that only 359,762 signatures were legitimate[59] and that organisers had failed to get enough signatures (5% of the electoral role) to trigger a national referendum.[60] In response, Yasunidos accused the government of fraud, a claim which the government rejected. On 3 October 2014, the National Assembly authorised drilling in block 43 but made it conditional on the fulfilment of certain standards, minimising the environmental impact and effects on the indigenous peoples living in the area.[61]

Much has been written about the Yasuní-ITT Initiative and the reasons that ultimately led to its failure, including questions surrounding the financial and project-related implementation mechanisms,[62] its institutional weakness on the domestic level,[63] and Correa's own mismanagement of, and commitment to the Initiative. Socially and environmentally at stake in the Initiative was the cultural survival of the local Tagaeri and Taromenane indigenous communities impacted by drilling in the ITT oil fields, and the conservation value of the Yasuní National Park as the most biodiverse hotspot in the western hemisphere.[64] Yet economic interests in the Yasuní-ITT oilfields amount to billions of dollars in increased GDP growth in Ecuador, seen as the platform for driving development and poverty reduction. Correa's decision to exploit the oil reserves in block 43 reveals the pragmatic calculations of traditionally leftist presidents prioritising economic growth through natural resource extraction over the social and environmental pillars of sustainable development.

Drilling in the ITT oil fields has brought unprecedented human access to one of the most intact portions of the Ecuadorian Amazon,[65] amounting to 'irreversible' consequences for ecosystems which are already threatened.[66] However, in July 2016 it was discovered that the ITT oil fields in block 43 hold at least 750 million additional barrels of oil, 82% more than originally certified. Correa wrote in his Twitter account that 'in current prices, this represents more than $26,000 bn and an increase in production of 300,000 bpd in 2022', further stating that 'Our best decision was to develop the ITT field.'[67] The hallmarks of socio-environmental conflicts surrounding natural resource exploitation, including lack of public consultation and participation, threats to local livelihoods and to indigenous rights, ecological justice and human rights,[68] are evident in the cancellation of the Yasuní-ITT Initiative and Correa's decision to exploit the ITT oil fields. In this sense the Initiative, and its subsequent cancellation are evidence of political moves to the left that require pragmatic steps that are inherently contradictory and inevitably lead to conflict.[69]

## 2. Sustainable development and human rights

In recent years human rights have assumed a central position in the discourse surrounding international development.[70] The outcome document of the 2012 United Nations Rio+20 World Summit committed member states of the UN to develop a post-2015 sustainable development agenda that identified poverty eradication 'as the greatest global challenge facing the world today and an indispensable requirement for sustainable development'.[71] According to a high-level roundtable discussion at the 29th session of the UN Human Rights Council, the whole message of sustainable development, as contained in the SDGs, is the integration of human rights and development.[72]

At the centre of the 2030 Agenda for Sustainable Development (Agenda 2030) are 17 S-Gs and 169 targets that seek to 'stimulate action over the next 15 years in areas of critical importance for people, planet and prosperity'. Human rights are at the core of the entire range of goals and targets,[73] and if not explicitly, a human rights-based approach is implicit in the agenda.[74] The SDGs aim to protect the planet from degradation through sustainable consumption and production, sustainably managing natural resources and taking urgent action on climate change.[75] They provide a plan of action for ending poverty and hunger and a roadmap for building a life of dignity for all, promising to leave no one behind.[76]

Additionally, the Paris Agreement on Climate Change (Paris Agreement) is the first environmental treaty calling on states to respect and promote their human rights obligations in addressing climate change, and to acknowledge that the right to health and the rights of indigenous peoples, migrants, children and persons with disabilities, among others, are particularly affected by climate change.[77] However, according to UNSG Ban Ki Moon 'the fact that 177 States have signed the Paris Agreement in less than a month is very welcome news, but the hard work of safeguarding the environment and human rights is just now beginning'.[78]

The term 'sustainable development' was launched in the UN-commissioned Brundtland report, *Our Common Future* (1987). It defined sustainable development as 'development that meets the needs of the present without compromising the ability of future generations to meet their own needs' and signalled a change in which issues of environment and econ-omic growth would be considered together.[79] To a significant extent, and because of the ambiguity of its definition, sustainable development allowed different concerns and inter-ests to meet, and at the 1992 UN Earth Summit in Rio de Janeiro the idea that economic growth can be reconciled with environmental conservation gained wide support from countries in the north and south.[80] Whereas development had long been discredited due to its association with foreign development cooperation and capitalist mega-projects with little sensitivity to the needs and livelihoods of local populations and vulnerable environments,[81] now no longer seen as an environmental threat or cause of global inequality, development becomes the route to sustainability.[82] In this way, governments around the world could circumvent discussions of the politically challenging issues necessary to reduce poverty, increase equity and create more environmentally friendly ways of living. However, the tensions surrounding these issues cannot be evaded by 'stimulative definitions'.[83]

A common vision of sustainable development envisages a mixed economy in which states and markets work in harmony to sustainable ends. UNSG Ban Ki Moon has affirmed that 'the private sector is the engine that will drive the climate solutions we need to reduce climate risks, end energy poverty and create a safer, more prosperous future for this and future generations … '.[84] Addressing the international business community at the World Economic Forum at Davos in January 2016, Mr Ban highlighted that it is governments that must take the lead in implementing the landmark global agreements contained in the SDGs and the Paris Agreement.[85] 'At the same time, businesses can provide essential solutions and resources that put our world on a more sustainable path.'[86] However, while research centred on energy efficiency and exploration sustains the notion that solutions lie in technical innovation, energy resources are converted into political power in complex ways[87] and there are a wide range of important critiques, including matters of fundamental political, economic and social concern, which identify many aspects of human rights praxis as being deeply problematic in the context of processes of accelerated globalisation.[88]

The overwhelming importance of commodity exports to Latin American economic development – linked to the 'resource curse' thesis[89] and Karl's 'cycle of inequality'[90] – reflects the growing external demand of developed countries to meet energy consumption needs in the context of continuous global growth mainly driven by the increase in population.[91] However, critiques of the resource curse thesis[92] link it to a state-centric neoliberal focus on the globalisation of environmental policy-making that has been accused of 'flattening uneven relations of power'. Critics assert that such a focus minimises the role of different non-state actors (including international investors such as China or International Financial Institutions (IFIs), indigenous groups, social movement coalitions such as Yasunidos, international NGOs and funders, and international organisations such as the UN) and their participation in socio-environmental conflicts and the production of socio-environmental order.[93]

Within a state-centric system of international governance that promotes a universal acceptance that there can be no development without economic growth, and no economic growth without free trade, the protection of economic and social rights also becomes difficult and raises questions regarding the responsibility of private and economic actors for both the promotion of human rights, and addressing rights violations.[94] As governments compete for corporate investments by participating in a 'race-to-the-bottom' that includes reducing social protections, business in the globalised world circumvents the shackles of regulation to seek greater profit margins and shareholder returns.[95] Within a concept of sustainable development that encompasses both a human rights/human development ideology on the one hand and a good governance ideology that encourages private investment for economic development on the other, this tension between the human rights obligations of companies and laws binding states is not resolved.

The degree to which environmental considerations and governance mechanisms are incorporated into state development strategies depends not only on the development priorities of the state, but also on their success in enrolling or competing with these alternative elites.[96] If openness to trade and investment is essential to development, rich countries have more power to protect their industries and dominate multilateral trading bodies such as the World Trade Organisation,[97] where concerns about competitiveness are increasingly being used by foreign corporations employing clauses written into free trade agreements to challenge restrictions posed by national environmental regulations on their activity. This raises questions about the sovereign power of Latin American nation states to determine how they will pursue strategies of sustainable development.[98]

Analysis shows that 30 years of historical data, including issues such as peak oil, climate change and food and water security, resonate strongly with the feedback dynamics of 'overshoot and collapse' displayed in the *Limits to Growth*[99] 'standard run' scenario,[100] which results in collapse of the global system midway through the twenty-first century. The key driver behind the *Limits to Growth* prediction, and arguably the one most poised to quickly cause global economic collapse, is the depletion of non-renewable energy sources, especially oil and natural gas.[101] Studies suggest the Latin America is likely to be disproportionately affected by climate change[102] due to its dependence on extractive and export sectors, its climate variability (particularly the El Nino phenomenon which already causes adverse impacts on many countries) and its inequality and poverty rates which climate change will exacerbate.[103]

Effective responses to climate change will need to differ based on local circumstances and adopt a plural notion of development that is neither rigidly statist nor based solely on the free market.[104] However, the failure of the Yasuní-ITT Initiative supports the pragmatic argument that to be fast, effective and politically acceptable, development and environmental approaches need to develop strategies that work with the economic interest mechanisms of the neoliberal framework of industrialised countries.[105]

## 3. Neo-extractivism in Ecuador

Despite its relatively small size (283,560 km$^2$), Ecuador is the country with the largest absolute area covered by oil blocks in extraction in the western Amazon basin. Operative oil blocks already occupy one-third (32%) of the Ecuadorian Amazon, and with the end of this bidding process of the XI Ronda Petrolera, Ecuador will have the majority of its Amazon, 68% (68,196 km$^2$), compromised by oil operations.[106] Initiated by Texaco Gulf, oil extraction began in the Ecuadorian Amazon (known as the Oriente) in 1967 and oil exportation began in 1972, since when the petroleum industry has been Ecuador's largest contributor of economic growth.[107] The smallest member of OPEC, since the 1970s Ecuador's percentage of daily crude oil production per day has increased at a faster rate than OPEC overall.[108] In 2012 petroleum exports represented nearly 60% of all exports and over 10% of national GDP, and have represented approximately 30% of central government revenues for most of the past decade.[109]

In line with the neo-extractivist policies of the populist, left-turn governments of Latin America (including Venezuela and Bolivia), revenues from hydrocarbon production are essential to financing the public budget in Ecuador.[110] Neo-extractivism[111] allows the state to exercise more centralised power with the aim of filling the state coffers to enable distribution and poverty reduction (among other aims), and to promote policies intended to compensate for the negative environmental and social effects of natural resource exploitation. Since coming to power in 2007 President Correa's *Revolucion Ciudadana* (citizen's revolution) has increased government spending on public sector expenses and investment to levels not seen in decades,[112] from 21% of the GDP in 2006 to 44% in 2013.[113] This includes doubling poverty assistance payments and credits available for housing loans, subsidising electricity rates for low-income consumers, and re-channelling millions of dollars into social programmes.[114] Investment programmes in energy, infrastructure and transportation, as well as in social sectors have led poverty (measured by income using the national poverty line) to decrease from 37.6% to 22.5% between 2006 and 2014, and extreme poverty to reduce from 16.9% to 7.7% in the same time period. In addition, the Gini coefficient was reduced from 54 to 48.7 between 2006 and June 2014,[115] signalling that inequality reduction has been quicker than the region's average as growth benefitted the poorest.

Correa's promise to 'refound' the Ecuadorian state and to reclaim national sovereignty back from global actors was widely considered to be a popular, if also populist, mandate.[116] Using environmental issues as an integral part of the discourse of change,[117] Correa framed his movement as one in which people of the developing world 'become the owners of our countries, the owners of our democracies'.[118] The 2008 constitution, passed in a popular referendum in September 2008, created the National Decentralised Participatory Planning System (SNDPP),[119] endowing it with the power and resources to develop and implement a vision for development founded on the concept of *buen vivir* or the *sumak kawsay*[120] in kichwa. As the orienting concept of the new Ecuadorian constitution,[121] *buen vivir* requires that 'persons, [indigenous] nationalities and peoples effectively enjoy their rights and exercise responsibilities in the frame of interculturality, respect for diversities, and harmonic co-existence with nature'.[122] Eduardo Gudynas, consultant to the Constituent Assembly for the 2008 constitution, describes *buen vivir* as 'a plural concept designed to be an alternative to western models of development … it incorporates a set of rights that include, among others, those of freedom, participation, communities, protection, and the rights of Nature[123] plus the rights to the resources found on indigenous territories and the exercise of the right to control and manage those resources'.[124]

The *Plan Nacional Para el Buen Vivir 2009–2013* (national development plan) complied with the principles of a model of sustainable development. Published early in President Correa's administration, the plan was based on three pillars: economic, political and environmental. At the economic level, the government outlined its aims to foster sustainable economic growth through selective import substitution, diversification and redistribution in key productive sectors. The political pillar is the establishment of a system of decentralised management for territorial planning and capacity building. At the environmental level, it aims to guarantee the sustainable exploitation and protection of natural resources.[125] The second *Plan Nacional* (2013–2017) acknowledges that 'the existence of oil camps brings opportunities to generate income. However, it also recognises that the socio-environmental impacts of oil extraction are very high, such as settling protected lands, deforestation, and the resulting habitat degradation, loss of biodiversity, contamination of soils and water sources, and others.[126] It specifically calls for diversifying national production away from oil by expanding renewable energy sources, promoting energy efficiency and moving towards a 'post-petroleum economy' based on tourism, eco-tourism, scientific research, and other services linked to the conservation of biodiversity.[127]

A post-petroleum economy provides for the significant participation of megadiverse countries and indigenous communities in the results of future scientific research, generating sustainable employment that is responsive to investments in education, health and human development. Economic diversification, biodiversity conservation and social equity will complement each other in a new path towards a sustainable society.[128] In this sense the government's adoption of the Yasuní-ITT Initiative in April 2007 was not only part of a national environmental policy but also a key ingredient of a new national energy policy that includes the oil sector but places it in another context alongside diversified sources of energy.[129] The pursuit of neo-extractivism through state-led capitalist development[130] creates a strong tension between the means and goals of Ecuador's development policies, if the goal is a new environmental paradigm and public reorientation towards a post-petroleum economy.[131] These divisions are apparent in the contradictions that are marking the progress of almost the entirety of government policy with regard to oil production, mining and agribusiness in Ecuador.[132]

State-building projects are pursued in line with the development model promoted by the Integración de la Infraestructura Regional Suramericana (IIRSA), that promotes access roads and railways in a continent-wide push to open up frontiers for extracting hydrocarbons, mining, producing biofuels, harvesting timber and investing in agro-industry.[133] These include policies such as new mining and water legislation, aimed to encourage foreign investment in its extractive industries, but which continues to alienate those groups that have been most affected by its doctrine.[134] The Correa administration has consistently ranked among the most popular in Latin America,[135] with Correa and other members of his Alianza País party easily winning the presidency and maintaining strong control of the National Assembly in February 2013. Yet Correa's popularity has been volatile.[136] Many of Ecuador's civil society organisations focused on human, environmental and indigenous people's rights are now opposed to the president[137] and in early 2014 elections the party lost mayoralties in four major cities: Quito, Guayaquil, Cuenca and Manta. Many Ecuadorean pundits connected this loss to Correa's decision to exploit oil in the ITT section of Yasuní National Park and targeting of indigenous and environmental critics of his policies.[138]

## 4. Case study: oil exploration in southern Ecuadorian Amazon

Contracts for the exploitation of oil fields in Ecuador involve the concession of delimited blocks which have a maximum surface area of 200,000 ha.[139] In 2011, the Ministry of Hydrocarbons produced the most recent restructuring of the Ecuadorian oil map, taking the total number of oil blocks to 65, with 57 of them in the Amazon. The *XI Ronda Petrolera* bidding round, which began in November 2012, has opened up the 21 new blocks in the southern Amazon for oil concessions. To date, offers for only four oil blocks (28, 29, 79 and 83) have been submitted by oil companies. In January 2016 Ecuadorian Minister of Hydrocarbons Carlos Pareja affirmed that by mid-year a new oil round will be opened with the aim of expanding exploration in the south-west to increase oil reserves.[140] At the end of these negotiations, all blocks not allocated to private or mixed companies will be assigned to Petroamazonas.[141]

In January 2016 the Ecuadorian government signed two contracts with Andes Petroleum Ecuador (Andes Petroleum), a consortium of two Chinese state-owned firms – China National Petroleum Corporation and China Petroleum and Chemical Corporation – to work on exploiting oil blocks 79 and 83 located in the Amazonian province of Pastaza. The new $72 m deal establishes a four-year exploration period, followed by a 20-year period of drilling.[142] According to Ecuadorian Minister Rafael Poveda (Coordinator of Strategic Sectors), after studies to identify possible reserves, it will be estimated whether exploitation is feasible or not, and if it is, exploitation and income production for the state would start.[143] According to Zhao Xinjun, the President of Andes Petroleum, the consortium has invested $3.5 bn in Ecuador 'with cutting-edge technologies and full respect for the country's environmental regulations[144] ... Cooperation with Ecuador is based on a winning relationship generating mutual benefits'.[145] For Ecuadorian Minister of Hydrocarbons Carlos Pareja, the deal 'sends a message that our country is building up confidence and that companies want to come and invest here despite the low international oil prices'.[146]

However, conflict and controversy surround the contracts to exploit the oil reserves in blocks 79 and 83. While Ecuador has established a top-level legal framework to limit the social, environmental and economic risks associated with the oil industry, this framework is facing increasing pressure from a diverse array of interests. Furthermore, this expansion is the first new concession under Ecuador's recent law on prior consultation, and problems have already arisen in how the law has been applied.[147] According to the International Working Group on Indigenous Affairs (IWGIA), the number of conflicts related to oil and mining in the south-east Amazon and southern Andean regions of Ecuador numbers on average 50–80 per month.[148] The analysis below uses the dynamics of the conflict surrounding the expansion of oil exploration in blocks 79 and 83 in the southern Ecuadorian Amazon as a lens to explore key tensions between the three pillars of sustainable development. Much of the tension between indigenous and environmental organisations and the state in Ecuador has concerned the role of nature within a development model that could be a realistic alternative to neoliberal capitalism.[149] The roles of all actors in the conflict are explored, including Andes petroleum, the Ecuadorian state, indigenous peoples and environmental groups. Also included are references to the conflict surrounding the exploitation of the Yasuní-ITT oil field (block 43) and Ecuador's obligations in international human rights law.

### A: Economic interests

Ecuador's 2008 constitution stresses the government's control over extractive industries, defining the oil sector as a strategic asset and declaring the state's inalienable ownership. In this 'post-neoliberal' model, the state, not the foreign firm, extracts oil with the aim of using oil rents to support social services and President Correa often publicly states the centrality of a state-run oil industry to a post-neoliberal transition.[150] In defending government policies against opponents such as Acosta, who referred to petroleum as a resource curse, Correa has maintained that creating alternatives to an extractive economy was a long-term proposition, short-term dependence on mining and oil extraction for revenue and employment was unavoidable and that he is determined to use Ecuador's natural resources to create a positive development model.[151]

Starting in 2007, the Correa government introduced reforms to increase the Executive's power over the national budget. In 2008, it enacted a law that eliminated all oil funds and their earmarked expenditure. When Correa eliminated the funds, these $2,000 m in savings, representing around 3% of GDP, became available and entered the budget. In 2007 Correa's government also amended the 2006 Windfall Tax law – under which additional revenue from any increase in the price of oil over the contracted price would be split equally between the government and the oil companies – to split the distribution of additional oil funds 99–1 in favour of the government, and then finally at 70–30 in 2008.[152] This helped to increase tax revenue from 5.6% of GDP in 1996 to 14.5% by 2012, which has funded redistribution through direct transfers and universalism.[153]

With more resources and greater flexibility, the government was able to increase expenditure per capita from $146 in 2006 to $297 in 2012 and also invest in infrastructure projects.[154] The Correa administration established the Gobiernos Autónomos Decentralizados (GAD) to boost devolution of oil revenues to sub-national governments at the province, canton, and parish level. Over the decade since the fund was established, Ecuador has produced nearly two billion barrels of oil, distributing the resulting funds to municipalities (58%), provinces (28%), parishes (5%), and the ECORAE (9%). The funds come with restrictions: at least 80% must be spent on conservation and transportation projects, and the rest is to be spent on public investments approved by the Secretary of Hydrocarbons.[155] More recently, the 2010 oil reform law redirected 12% of oil profits (which previously went to the central government) to the GADs in the regions where the drilling takes place, to be used for health and education projects as approved by the appropriate ministry.[156]

This is reflected in Ecuador's considerable reductions in inequality since 2006, achieving the world's most 'inclusive' growth if we consider the growth rate of the incomes of the poorest 40% of the population relative to the average: in Ecuador these grew over eight times the rate of the average. Ecuador's Palma ratio[157] also fell by half between 1999 and 2012, from 5.2 to 2.6. According to the Unmet Basic Needs indicator of multidimensional poverty (*Necesidades Basicas Insatisfechas* – NBI), multidimensional poverty fell from 65% to 52% between 1998 and 2006, and to 36% by 2014.[158] However, oil remains Ecuador's most important export and any fall-off in oil production can represent a major disruption in local government finance so the goal of diversification remains an important priority.[159]

Part of Correra's strategy to buffer Ecuador against the volatility of overwhelming reliance on the US as an export market has involved looking to China as an investment partner and favouring socially responsible large-scale mining and infrastructure operations governed by strong state control to protect the environment and workers' rights, while also emphasising the necessity of access to the revenues that resource extraction would generate to fund important social programmes.[160] Ecuador's 'China boom'[161] encompasses investment, finance and trade, and is overwhelmingly concentrated in the oil sector.

China is now Ecuador's most important creditor (accounting for over one-third of the nation's total external public debt in 2013) and has guaranteed Ecuador access to financial markets.[162] Chinese investors have played a crucial role in the development of Ecuadorean oil fields and have strengthened the position of the Ministry of Non-Renewable Resources in Ecuador, as new sources of funding for energy and infrastructure projects enter the government coffers.[163]

According to documents examined by Reuters,[164] in November 2013 PetroEcuador, the state-owned oil company, signed an agreement with the Chinese state-owned company PetroChina, in which Ecuador undertook to sell over 90% of its oil production to China until 2020. This suggested that China was close to monopoly control of crude oil exports in Ecuador. In July 2009 PetroChina had deposited $1 bn in the coffers of the Ecuadorian government with an agreement that they would be returned over a two-year period at an interest rate of 7.25% (very high) with a concession of 96,000 barrels of crude oil per day to Chinese companies. That same year, which also coincided with when the Yasuní-ITT Initiative was officially launched, the Ecuadorian Ministry of Economic Policy in a private presentation to Correa's staff, pledged to 'make the utmost effort to support Petro-China and AndesPetroleum in the exploration of the ITT oilfield'.[165] According to Reuters, official documents of an opening of a $1 bn credit line from the China Development Bank in 201, reveal the Correa government gave PetroChina permission to resell Ecuadorian crude oil in any market.[166]

In January 2016, around the same time as the contracts were signed for oil blocks 79 and 83 in the southern Amazon, Ecuador agreed a $970 m credit line with the Industrial and Commercial Bank of China (ICBC).[167] Critics of Correa's government, including Amazon Watch, claim that this latest deal is evidence of the 'millions of dollars of debt that Ecuador is having to pay back [to China, which is] forcing the expansion of the oil frontier in the Ecuadorian Amazon. These investments are moving forward because a lot of the conditions within these contracts are tied to future oil sales'.[168]

Ecuador is heavily exposed to debt, particularly China's. Excluding the recent $970m loan, Ecuador currently owes its partner $5.4bn.[169]

The Correa administration has planned to use the perceived revenues of expanding the oil industry in southern Ecuador for further investments towards economic diversification.[170] However, by tying their financial support to extractive industries, particularly petroleum, China limits Ecuador's ability to diversify production and pursue a vision of sustainable development that protects its ecosystems.[171] High oil revenues produce institutional rigidity and high barriers to reform.[172] From 2008 to 2012, extractive products made up 58% of all Ecuadoran exports, but 70% of exports to China.[173] That institutional rigidity increases as oil prices increase is the essence of the 'paradox of plenty'[174] or the resource curse thesis discussed above. The trade balance disadvantage of indebted countries with natural resources that are in high demand, poses a huge challenge to countries aiming to diversify their economies away from a reliance on petroleum exports.[175] As the *Plan Nacional* states, in an economy that is based on 'extraction and export of commodities, long-term economic growth revolves around external market dynamics, especially the price of oil,[176] and neglects internal demand ... to the detriment of national production and employment'.[177] The impact of commodity price volatility on aggregate demand, savings and investments and output rises with the degree of export concentration.[178]

Divestment efforts to date have included micro-loans with preferential terms and special attention to the non-petroleum sectors. From 2007 to 2012, the National Development Bank and the Ministry of Economic and Social Inclusion issued more than a million small loans, totalling nearly $3 bn (about 0.4% of GDP), to individuals and small businesses. Furthermore, infrastructure and education projects provide support for businesses of all sectors, and can help support competitiveness in non-traditional industries. However, as beneficial as these programmes may prove to be in the long term, they have not proven sufficient to reduce Ecuador's dependence on oil.[179]

In a further effort to diversify the economy, Correa has pledged to make mining a priority in his third term and is actively courting international expertise in sectors such as mining, energy and infrastructure.[180] Correa favours socially responsible large-scale mining and infrastructure operations governed by strong state control to protect the environment and workers' rights, while also emphasising the necessity of access to the revenues that resource extraction would generate to fund important social programmes.[181] Reforms in 2003 and 2007 require environmental impact assessments (EIAs) for new projects (including plans for prevention, remediation and compensation of contamination) and earmark large shares of oil revenues for local governments in areas affected by the oil industry, to fund public investment projects approved by the central government. Yet the redistributive policies that have been implemented relate mainly to the sphere of social services, while the economic and the environmental areas – where structural change would demand redistribution of assets and reforms in the management of productive resources – have not been significantly addressed.[182]

## B: Environmental interests

There is extensive documentation of the direct and indirect environmental impacts caused by oil exploitation in the Ecuadorian Amazon.[183] In practice, oil activity has been characterised by the use of obsolete technology, the application of poor environmental controls[184] and the opening of roads in the rainforest which drives widespread colonisation, deforestation and agriculture expansion.[185] Exploration activities, road openings, noise from platforms and spills of oil and toxic waste into freshwater systems, may also affect large areas beyond wells and camps.[186] Beyond doubt, these and other environmental impacts caused by the oil industry are reducing biodiversity and threatening wildlife.[187] Contact with intercultural settlers has also changed indigenous culture patterns, increasing commercial hunting of wildlife to levels that could compromise ecosystem functioning. These, combined with the historical negative impacts of the oil industry in the environment of the Ecuadorian Amazon, have led to an understanding that oil blocks can be considered as spatial units where the environment, biodiversity and the health of local populations are potentially vulnerable.[188]

Ecuador's 2008 constitution became the first in the word to recognise rights for nature and 'integral respect for its existence and for the maintenance and regeneration of its life cycles, structures, functions and evolutionary processes'.[189] It granted strong rights to the state to control and nationalise resource industries[190] and banned resource extraction in protected areas altogether.[191] However, the constitution contains a clause that 'resources may be exploited [in protected areas] by request from the Presidency of the Republic and following a declaration of national interest by the National Assembly'.[192] As a result of this policy more than 30% of the protected areas coincide with operative oil blocks and only 16% of the Ecuadorian Amazon is actually covered by protected zones free of oil concessions.[193] Therefore, environmental scientists and conservation practitioners have concluded that this current network of Protected Areas cannot adequately protect the diversity of species and ecosystems against the development of the oil industry and other human threats.[194]

Confirmed oil reserves in the southern Amazon are relatively small in comparison to those in the northern Amazon, where operative oil blocks are concentrated among the highest species richness areas in the Ecuadorian Amazon for amphibians, birds, mammals, butterflies and vascular plants.[195] However, the expansion and intensification of oil exploration into the southern Ecuadorian Amazon seriously increases the vulnerability of peoples, species and ecosystems[196] because it encompasses 30,000 km$^2$ of virgin forest, holds outstanding biodiversity and is home to numerous indigenous communities of different nationalities.[197] Furthermore, given the limited conservation options in the northern Amazon because of the long-established oil exploitation, preserving the southern Amazon becomes essential to improve the protection of Amazonian species and ecosystems.[198] The southern Ecuadorian Amazon is especially vulnerable to biodiversity losses because peaks of species diversity, 19 ecosystems, pristine forest and a third of its protected zones coincide spatially with oil blocks. The cultural diversity of the peoples inhabiting the region and number of species and ecosystems distributed only in the southern Amazon make these areas irreplaceable.[199]

Chinese oil companies CNPC and Sinopec operate in Ecuador under the names Andes Petroleum and PetroOriental. They are two of the most successful foreign oil companies in Ecuador and are among the most important oil producers in the country. Taken together, they account for about a quarter of Ecuador's total oil production. Andes Petroleum alone produces more oil than any other private producer except for Repsol[200] and it also has a better environmental record, with fewer local protests over spills, than most of its foreign or domestic competitors in Ecuador.[201] The comparatively peaceful company-community relationship enjoyed by Andes Petroleum previously may be due to the location of its northern concession (block 62) in Sucumbíos, which has been home to large-scale agricultural and oil development, including Texaco's original oil fields, for decades. Within the controversy surrounding Texaco's toxic legacy in the Ecuadorian Amazon,[202] Andes Petroleum's ability to maintain a low profile has been key to its ability to continue operations for nearly a decade.[203] However, the economy and ecology of the new concessions in blocks 79 and 83 are quite different from Andes Petroleum's current concession in the northern region of Tarapoa, Sucumbíos. And unlike the Tarapoa concession, the new blocks will be greenfield projects[204] and this will be the first time the company has established new concessions anywhere in Ecuador.[205] Because of these differences, it seems very unlikely that Andes Petroleum will be able to expand its operations with the same positive community and government relations it has enjoyed in the past.[206]

Leading biologists working in the Ecuadorian Amazon have consistently argued that roads are the biggest threat and that the Ministry of Environment is not equipped to control accessibility in these remote sections of the forest. Analysts often point to block 16, which overlaps the Yasuní National Park, as an example of best practice for drilling with limited road construction. US company Maxus built a road in block 16 (now run by Repsol) from their operations to a nearby river, but left it unconnected from the country's highway system to discourage the development of new towns in the park. Instead, equipment and trucks must use barges to reach the road and eventually the oil installations.[207] In 2012 Repsol sold a 20% stake in block 16 to Tiptop Energy, a subsidiary of Sinopec (joint owner of Andes Petroleum), so Andes Petroleum does have privileged access to the technology used in Repsol's lower impact methods.[208] However, even the limited use of roads in block 16 has had a 'significant' impact on the Yasuní National Park, leading to deforestation, migration and unsustainable hunting practices.[209] Proximity to the Via Maxus access road is the strongest spatial factor in predicting where continued deforestation is occurring[210] at an annual rate of 0.11%. At that rate, by 2063, 50% of the forest within 2 km of the road will be deforested due to human settlement and forest deterioration.[211] To truly address road-based deforestation, Andes Petroleum would need to attempt the 'offshore-inland' model promoted by the Blue Moon Fund (BMF), in which roads are severely limited or not built at all, and equipment is brought in by helicopter.[212]

The prospect of new oil exploration in blocks 79 and 83 has brought intense criticism from environmental and indigenous groups, including Amazon Watch, the Pachamama Alliance, Acción Ecológica, and others.[213] These new blocks are just outside of the Yasuní National Park, however the area actually boasts a higher level of biodiversity than the park itself. Most of the area covered by the new concessions has extremely high biodiversity in four major categories: amphibians, birds, mammals and plants.[214] In contrast, most of the park has high biodiversity in two or three of those categories. Thus, from a conservation standpoint it is arguably at least as important to treat the ecosystem in blocks 79 and 83 carefully as in the park itself.[215] Many experts on the Ecuadorean Amazon, such as biologist Santiago Espinosa and conservationist Kelly Swing, believe that the government currently lacks the institutional capacity to successfully manage the ecosystems near the planned extraction sites.[216] Federico Auquilla, former Vice Minister of Mines and current advisor to Chinese firms in Ecuador, corroborates this view, though he also expressed confidence that the government will be ready for these responsibilities by the time the new greenfield projects come online.[217]

While Chinese social and environmental safeguards for outbound investment are groundbreaking for a middle-income country they still lag behind those of the traditional multilateral lenders, lacking important enforcement power and transparency.[218] Chinese investors in LAC have shown an ability to exceed local standards and have also enacted guidelines for their overseas economic activities,[219] but the performance of Chinese investors varies widely across different regulatory regimes and between more experienced and newer firms.[220] That Chinese banks are not on par with the environmental guidelines of Western banks is of 'grave concern' given that the composition and volume of Chinese loans is potentially more environmentally degrading than Western banks' loan portfolios to LAC[221] and also has a disproportionate climate impact. LAC–China exports create approximately 15% more net greenhouse gas emissions per dollar than other exports, and over twice as much as overall economic output.[222]

### C: Social interests

Unlike the northern Tarapoa concession (block 62) , the new southern concessions in blocks 79 and 83 are entirely covered by traditional indigenous territory and lie within the parish of Montalvo, Pastaza, among the poorest in the country.[223] Oil exploration and extraction necessarily impinge on the surface uses of those people who live on the same lands or territories, not only through the mode of extraction itself but also through the changing social character of life on the surface.[224] The close relationship between indigenous peoples and *Pacha Mama* (Mother Earth) is based on a duality and complementarity between all elements that make up the indigenous world view.[225] It entails living in balanced harmony with nature, recognising environmental limits and fostering community participation.[226] It is a world view that does not recognise commercial values or trading in natural resources such as land, water, minerals and plant life.[227]

• Sovereignty over territory and natural resources

The collective rights of indigenous peoples are guaranteed in Ecuador's 2008 constitution, especially the rights to maintain, develop and strengthen their identity and traditions, not to be displaced from their lands and to the protection of ritual and sacred sites and ecosystems.[228] International law has developed a clear principle of the right of indigenous peoples to permanent sovereignty over natural resources, based on the principle of self-determination contained in common article 1, paragraph 2 of the two International Covenants on Human Rights and on the United Nations Declaration on the Rights of Indigenous Peoples (DRIPS).[229] The principle of permanent sovereignty is an integral part of the right of self-determination, including the right to participate in the governance of the state and the right to various forms of autonomy and self-governance.[230] As detailed above, the clause in Ecuador's constitution allowing for resources to be exploited in protected areas following a declaration of national interest by the National Assembly translates into practice as only 16% of the Ecuadorian Amazon is covered by protected zones and free of oil blocks. Therefore, despite the impressive language in Ecuador's constitution, in practice Ecuador's oil activity in protected areas is a threat to the rights of indigenous peoples to permanent sovereignty over natural resources and self-determination, as protected in international law.

In interviews carried out in 2014 with then Sápara President Klever Ruiz and President of the Association of Sápara Women Gloria Ushigua, both indicated that the community was deeply divided between ethnic Sápara, opposed to the drilling, and other residents who are in the minority but who welcome the oil exploration.[231] When former Sápara President Mucushigua (who signed the agreement allowing oil exploration blocks 79 and 83) was asked about the agreement, he threatened to have anyone who stood in its way killed. Within a week, the 13-year-old son of an opposition community leader was allegedly murdered. Suspecting that Mucushigua was behind the death, the ethnically Sápara majority of the Sápara community met and elected Ruiz as the new president.[232] According to Amazon Watch, for the Sápara the conflict that the *XI ronda Petrolera* has caused among their people is one of the most harmful aspects of the oil round, 'It is not just about the contamination and the loss of their sovereignty but also about the loss of harmony amongst community members'.[233] However, as lawyer for indigenous groups Mario Melo has stressed, the wishes of indigenous groups as indigenous nations and peoples in their practice of self-determination and who may or may not seek out petroleum activity must be respected.[234]

• Cultural survival

The oil concession in block 83 coincides with the titled lands of the Sápara nation whose population is less than 300 and whose language is 'critically endangered' with only nine speakers.[235] The Sápara language is one of just two Ecuadorean cultural practices included by UNESCO in the Representative List of Intangible Cultural Heritage of Humanity. UNESCO highlighted the Sápara oral culture as 'particularly rich as regards their understanding of the natural environment ... demonstrated by the abundance of their vocabulary for the flora and fauna and by their medicinal practices and knowledge of the medicinal plants of the forest'.[236] The Sápara fear oil exploitation will lead to widespread soil, groundwater and surface stream contamination.[237] This is a threat to the medicinal plants of the forest, and therefore to the survival of their traditional medicinal practices.

The Inter-American human rights system, in *Awas Tigni* v. *Nicaragua* (2001) concluded that access to traditional lands is 'a material and spiritual element which they must fully enjoy ... to preserve their cultural legacy and transmit it to future generations'.[238] Territory is 'a fundamental basis for the development of indigenous communities' culture, spiritual life, integrity and economic survival.[239] It encompasses the use and enjoyment of natural resources and is directly related, even a pre-requisite, to enjoyment of the rights to an existence under conditions of dignity ... '.[240] As mentioned previously, exploration activities can affect large areas beyond wells and camps, and oil blocks can be considered as spatial units where the environment, biodiversity, and the health of local populations are potentially vulnerable. Contamination of the soil, groundwater and surface stream would also have a serious impact on the health of the Sápara, which constitutes a threat to their survival.

Block 83 also borders the Yasuní National Park, a 3,800-square-mile area of jungle which is home to two indigenous tribes, the nomadic Tagaeri and the Taromenane, who have no contact with the outside world.[241] Peoples living in voluntary isolation are protected by the *zona intangible* and article 57 of the 2008 constitution that specifically mentions that the ancestral homelands of peoples in voluntary isolation are 'irreducible and untouchable, and off-limits to all extractive activities'. Furthermore, the state will 'adopt measures to ... ensure that they can remain in voluntary isolation, respect their self-determination and ensure that their rights are respected'.[242] Oil exploration in block 83 therefore threatens the right of voluntary isolation for the Tagaeri and the Taromenane and the cultural survival of the Sápara nation.

• Participation

Blocks 79 and 83 are the first new concessions in Ecuador agreed under a new citizen participation law, but opponents of oil exploration accuse the government of failing to adequately consult affected indigenous populations before granting the concessions.[243] Article 57 of Ecuador's constitution and the 2010 Citizens Participation Law enshrine the requirement that communities have the right to consultation before developments in their traditional territory. According to Minister Rafael Poveda, Coordinator of Strategic Sectors, the signing of the contracts for blocks 79 and 83 authenticates the process of prior consultation completed by Ecuador as part of the *XI Ronda Petrolera* launched in 2012. 'It was part of a prior consultation and perhaps a unique example in our country's history, new in the region and an example of how these socialisation processes are done with communities, nationalities and people living in nearby areas.'[244] However,

Ecuadorean law also requires the Secretary of Hydrocarbons (SHE) to seek majority approval within the affected community[245] but in this case SHE circumvented the obligation by only seeking the approval of the Sápara president at the time, Basilio Mucushigua, who signed an agreement allowing oil exploration in exchange for $2.4 m in local public investment.[246]

The requirement of majority opinion was not included in Executive Decree 1247, which directed this particular consultation. The decree allows for comments to be submitted either through community meetings or individually at local consultation offices, provided that the offices are extensively advertised through local press, government or community leaders. SHE opened temporary outreach offices in the affected area of the southern Amazon, and claims that 16,469 people participated in workshops or submitted comments – a number equal to about a quarter of the local adult indigenous population, or about an eighth of the total adult population in the new concession blocks.[247] Although numbers are disputed, what they do show is that SHE consulted only a small minority of the affected population.[248] Manari Ushigua, current president of the indigenous Sápara Nation of Ecuador, claims his community were not consulted about the contracts.

Ecuador's national laws do not make extractive activity subject to the consent of the affected communities on whose territories developments will take place.[249] However, as well as being enshrined in International Labour Organisation (ILO) Convention 169,[250] the Inter-American Court (IACtHR) previously established the standard for the need of states to 'actively consult' indigenous peoples about activities on their land in the case of *Saramaka* v. *Suriname* (2007),[251] further affirming in the judgement of *Sarayaku* v. *Ecuador* (2012)[252] that 'the safeguard of effective participation … must be understood to additionally require the free, prior, and informed consent (FPIC) [of indigenous peoples], in accordance with their traditions and customs'. *Sarayaku* v. *Ecuador* marked the first occasion where the Inter-American Court expressly asserted that the duty of states to consult indigenous peoples with the aim of obtaining FPIC is non-delegable.[253] Therefore, Ecuador's national law, even if applied properly is not in line with the principle of FPIC established by the Inter-American legal system, and further stated in DRIPS.[254]

At a press conference in November 2013, Franco Viteri, President of the Government of the First Nations of the Ecuadorian Amazon (GONOAE) and President Humberto Cholango of CONAIE (Ecuador's National Indigenous Confederation), warned that the *XI ronda petrolera* of bidding was unconstitutional and in violation of human rights as it was being conducted without the FPIC of indigenous peoples and nations.[255] During a press conference held in Quito after the contracts were agreed in January 2016, leaders from the Kichwa Sarayaku, Achuar, Shiwiar and Shuar Amazon communities showed their support and joined the Sápara resistance against the contracts signed by the government, alongside GONOAE and CONAIE, two of the largest indigenous confederations in the country. The Sápara say they are ready to follow the Sarayaku example and bring their case to national and international courts in order to avoid the drilling of blocks 79 and 83. 'The community is alert and ready to prevent the entrance of Andes Petroleum into their territory.'[256] However, as Mario Melo, a lawyer for indigenous groups, has stressed, 'if the Sápara come to agreements that respect their rights and if through those agreements they see greater opportunities [in oil exploration], then that is good. The problem arises when efforts are made to impose something on them … It is vital to respect those who say yes as well as those who say no'.[257]

• Jobs

If opposition to the oil exploration in blocks 79 and 83 can be overcome, the demand for public works and jobs in Montalvo is sure to be much greater than in Tarapoa, given the high poverty rate in the area. Furthermore, communities living in the southern Amazon do not have the same decades of experience negotiating with the government and foreign companies as their northern counterparts, and it would therefore be naïve to expect negotiations to go as smoothly as they have in the north.[258] Andes Petroleum have had their share of labour disputes and early on in their presence in Ecuador they faced community conflict over local job opportunities.[259] However, Ecuador has enacted a series of labour protections that form one of the most progressive packages of labour protection for the Ecuadorean petroleum sector in the Amazon region.[260] In 2008, Ecuador strictly curtailed the use of subcontracted labour, limiting it to complimentary work such as security and custodial services. The 2010 Hydrocarbon Law further boosted labour protections in the oil and gas sector, by requiring foreign investors to hire Ecuadorean workers for 95% of unskilled and 90% of skilled jobs.[261] Moreover, it required profit sharing with all employees, including contract workers. These protections have largely eliminated problems over the issue of local employment.[262]

Chinese Embassy Attaché Zach Chen states that Andes Petroleum and PetroOriental have established English as the primary working language in their Ecuadorean facilities, only hiring workers who speak it fluently. According to Mr Chen, this limits the pool of potential workers, raises their salaries, reduces turnover and improves morale. However, this policy also limits hires from the immediate vicinity, where schools are not able to teach students sufficient English. According to interviews with Shushufindi Mayor Édgar Silvestre and human rights advocate Wendy Obando, the language requirement contributes to local unemployment and underemployment and will continue to do so until local schools are able to meet it.[263] So while this problem has been addressed at the national level, it may continue to cause friction with the local community in the future.[264]

• Peaceful assembly and association

CONAIE and other indigenous groups have left Correa's coalition government, criticising its lack of commitment to implementing plurinational and intercultural measures in political discourse and practice.[265] The Yasuní-ITT Initiative was portrayed by the state as a key policy demonstrating its concern for indigenous livelihoods and well-being.

However, according to Miguel Guatemal, Vice-President of CONAIE, 'there is no dialogue, rather there is a direct imposition … we as the indigenous, have to accept everything that they say'.[266] From these assertions it would appear that Ecuador failed in its obligations to guarantee the participation of indigenous peoples in the planning, and also in the decision to cancel the initiative. This lack of democracy, coupled with the dismissal of the Yasunidos petition for a referendum on drilling the ITT block, fuelled a high degree of ongoing civil discontent in Ecuador.[267] Martin Carbonell, a spokesperson for Yasunidos, told *The Guardian*, 'People are aware that this has damaged democracy. This was the moment when people could say this is not a democratic government.'[268]

According to the special rapporteur on FOAA, social conflicts experienced in the context of natural resource exploitation are a stark demonstration of the severe consequences and counterproductive nature of the failure to provide any outlet for excluded groups to air their grievances.[269]

The government's response to the protests surrounding the cancellation of the Yasuní-ITT Initiative included blocking the streets on which protestors were marching, firing rubber bullets and beating peaceful protestors.[270] The special rapporteur holds the view that any interference with peaceful assemblies, including dispersal, 'should meet the strict tests of necessity and proportionality stipulated in international human rights standards',[271] including that the right to freedom of assembly cannot be limited based solely upon an assembly's message or content.[272] Furthermore, the special rapporteur links the restriction of these rights to questions regarding the right to participation, and 'how genuine consultation processes or decisions are, and how valid is the expression of free, prior and informed consent of affected parties'.[273]

The government response also fuelled a general mobilisation by civil society to denounce the increasing criminalisation of social protest, and to demand open political debates and protection of the right to dissent.[274] The Ecuadorian government used an attack on dignitaries attending the opening session of the *XI ronda petrolera* in Quito to press charges against some of the demonstrators, including a number of indigenous leaders who were present. Humberto Cholango denounced the criminalisation of social protest that had led to activists defending themselves against charges of terrorism and sabotage.[275] Also in response to the protests, the Ministry of the Environment decided, by means of Agreement No. 125,[276] to dissolve the Pachamama Foundation.[277] According to the special rapporteur on FOAA, when violent incidents occur within otherwise peaceful assemblies, authorities have a duty to distinguish between peaceful and non-peaceful demonstrators, take measures to de-escalate tensions and hold the violent individuals and not the organisers to account for their actions.[278] The special rapporteur has also acknowledged that legal mechanisms, used to curtail the work of civil society organisations and individuals engaged in defending rights in the context of natural resource exploitation are of concern because of the chilling effect the proceedings may have on the legitimate expression of dissent.[279]

The Ecuadorian government has also been criticised by international NGOs for its use of criminal defamation prosecutions, anti-terrorism laws and administrative sanctions against critical journalists, media outlets, NGOs and human rights advocates.[280] This pattern of aggression in the form of repressive legislation, harassment, violence and threats is used as a disciplinary measure to deter other communities from mobilising in support,[281] and is a trend across the world. As *The Guardian* reported in December 2014, the killing of José Isidro Tendetza Antún'san, an indigenous Shuar leader, days before a protest at the UN COP20 in Lima highlights the 'violence and harassment facing environmental activists in Ecuador'.[282] The special rapporteur on FOAA has deemed human rights defenders in the context of natural resource exploitation as the most at risk from attacks and reprisals.[283]

States and private actors in many cases portray environmental human rights defenders as criminals, characterising their opposition to 'important' projects as against national interests or anti-development,[284] something the special rapporteur has termed 'the demonisation of protesters and human rights defenders'.[285] This can be seen in President Correa's complaints about 'infantile environmentalists' creating obstacles to economic development in the midst of the conflict surrounding the decision to exploit the oil fields in block 43. He dismissed groups that opposed him as part of an 'infantile left' made up of 'fundamentalists' who could not see that creating alternatives to an extractive economy was a long-term proposition, and short-term dependence on extraction for revenue and employment was unavoidable.[286]

We cannot lose sight of the fact that the main objective of a country such as Ecuador is to eliminate poverty. And for that we need our natural resources. There are people here who seem ready to create more poverty but leave those resources in the ground ... That is criminal ... .[287]

## 5. Findings

'We cannot be beggars sitting on a sack of gold' is the mantra repeated regularly by President Rafael Correa since he took office in 2007. The implication is that Ecuador should take advantage of its natural resource base, including petroleum and mineral wealth, in order to fund social development and redistribution.[288] That Ecuador has seen a substantial reduction in poverty and inequality on the back of Correa's neo-extractivist policies is not disputed, and is evidence of a significant redistribution of the benefits of economic growth in favour of the poorest 40%, albeit from a starting point of high inequality.[289] However this developmentalist version of *buen vivir* departs significantly from the popular demands that brought the indigenous concept of *buen vivir* into the political sphere to push for increased participation for indigenous peoples, sovereignty over territory and natural resources and protection of the environment.

Public support for Correa's Alianza País government, in an alliance with environmentalists and indigenous leaders, enabled the writing of the new 2008 constitution[290] with a guarantee of the realisation of *buen vivir* or *sumak kawsay* at its central principle in the 'regimen of development'. The proposal to 'keep the oil in the ground' also came from the civil society support base that carried Correa into government in 2007. The Correa government adopted the proposal as the Yasuní-ITT Initiative in line with its development agenda, and then proceeded to propose it to the UN. In this way the Yasuní-ITT Initiative came to be seen as the embodiment of sustainable development in practice, and there were great expectations regarding their willingness and ability of the Correa administration to shift development strategies towards more sustainable and equitable policies that pay more attention to the concerns of local populations and indigenous peoples, than to the exigencies of global capital and domestic elites.[291]

*Sumak Kawsay* and *buen vivir* as presented in the Ecuadorian constitution and in the Yasuní-ITT Initiative, promoted the rights of nature, the rights of peoples to live within healthy environments, the values of environmental security and maintenance of world biodiversity and increased participation for previously excluded groups. The constitution has an explicit commitment to the realisation of social rights, collective citizenship and environmental rights. In its practical implementation it recognises two strategies: sustainability and innovation.[292]

The contradictions between the rhetoric of *buen vivir* and the neo-extractive development policies of President Correa mirror contradictions in the rhetoric of Agenda 2030 for social inclusion, environmental protection and sustainable economic growth, and the pragmatic reality of world politics dominated by economic interests. The economic interests involved in oil exploration in the southern Amazon amount to billions of dollars, both for the investment opportunities and access to oil for Andes Petroleum, and for the Correa government in securing increased GDP growth, seen as the platform for driving development and poverty reduction. Correa's pursuit of further oil exploration reflects the pragmatic argument that to be effective and politically acceptable, development and environmental approaches need to develop strategies that work with the economic interest mechanisms of the neoliberal framework of industrialised countries.[293]

Agenda 2030 affirms that the goals of combatting inequality, preserving the planet, creating sustained, inclusive and sustainable economic growth and fostering social inclusion are interdependent.[294] It recognises that social and economic development depend on the sustainable management of the planet's natural resources, and determines to conserve and sustainably use natural habitats to protect biodiversity, ecosystems and wildlife.[295] The concept of sustainable industrialisation prescribed in SDG 9 is intended to maximise the developmental impact of natural resources for inclusive and sustainable industrialisation and people-centred economies that increase productive capacities, productivity and productive employment and financial inclusion.[296] In practice, however, Ecuador is free to engage in the natural resource sector in an unsustainable manner because it is both politically acceptable at the international level, and in line with the political interests of the traditional oligarchy at the national level, reflecting the prevailing dominance of an economic model of development, upheld by business corporations and IFIs seeking to consolidate forms of neoliberal governance.[297] Furthermore, this faith in capitalism's long-term prospects for the progress of development and the enhancement of human rights is symptomatic of 'a distinct academic and popular denial of the *Limits to Growth* prediction'.[298]

Within the conflict surrounding the decision to exploit the reserves in Yasuní-ITT block 43 and further expansion of oil exploration in the southern Ecuadorian Amazon, the preservation of nature is central because it is simultaneously a source of economic value, as well as being a set of relations that underwrite the well-being of individuals and communities.[299] This tension surrounding the value of natural resources is at the heart of the tensions between the three pillars of sustainable development, and also reflects the challenge facing Ecuador to advance the three core areas of *buen vivir* (human development, harmony with nature and a knowledge-based economic model) and avoid the potential trade-offs among them.[300]

Analysis of human rights in practice gives equal weight to what the social theorist's eye sees and also what participants in human rights networks themselves tell us about the meanings and experiences of human rights as it relates to other forms of social practice.[301] Community leaders, NGO participants and social movement activists are 'knowledge brokers' who experience human rights discourse 'betwixt and between' and translate the promiscuous concept of sustainable development into practice, from local arenas up and also from international arenas down. These actors experience human rights discourse as a kind of legal or ethical liminality that can both empower the relatively powerless and also place them at a greater risk.[302] The extent to which the conflict surrounding oil exploration in the Ecuadorian Amazon adheres to the observations of multiple international special rapporteurs regarding the human rights risks of indigenous peoples and human rights defenders points to the unique and distinct nature of conflicts surrounding natural resource exploitation. Human rights defenders are powerful in that they serve as knowledge brokers between culturally distinct social worlds, but they are also vulnerable to manipulation and subversion by states and communities.[303]

Correra's government is not a government of social movements even though it has incorporated much of their agenda into its discourse and policies.[304] Correa has gradually eliminated the autonomy of indigenous institutions by incorporating them into various ministries centralised in the government, and in addition to undercutting existing organisational efforts, Correa has not used his executive power to create new spaces for grassroots social movements.[305] Correa's pursuit of neo-extractivist policies, the decision to exploit the oil reserves in the Yasuní-ITT block and to expand oil exploration into the southern Amazon point to the conclusion that social movements cannot ever achieve their transformative agenda without gaining control over governmental structures.[306]

While Agenda 2030 reflects the needs of indigenous peoples[307] as some of the most vulnerable[308] and recognises that natural resource depletion and adverse impacts of environmental degradation, including land degradation, freshwater scarcity and loss of biodiversity add to and exacerbate the list of challenges which humanity faces,[309] this analysis has highlighted the irreversible environmental and social consequences of oil exploration in the Ecuadorian Amazon. In practice, the government of Ecuador in its pursuit of economic development threatens to, or has violated its obligations in national and international law for environmental protections, guarantees of indigenous autonomy (including FPIC) and the participation of communities affected by natural resource exploitation.

## 6. Conclusion

This article has shown that governments with progressive constitutions, and promoting a development model with the potential to be a realistic alternative to neoliberal capitalism, are willing to override and marginalise the rights of their traditional supporters in the pursuit of economic growth. These same supporters are the social sectors who have not shared in the benefits of economic growth and whose survival and livelihoods are intertwined with the environmental and social pillars of development. Poverty rates remain high in Ecuador,[310] where the 2013 Human Development Index (HDI) of 0.711 is below the average of 0.735 for countries in the high human development group and below the average of 0.740 for countries in Latin America and the Caribbean.[311] For the government to achieve a further reduction of poverty, it will have to pay greater attention to social sectors that have not shared fully in the benefits of growth, especially those living in rural areas, in certain geographical regions and the Afro-Ecuadorian and indigenous populations.[312]

While Agenda 2030 reaffirms that every state has and shall freely exercise full permanent sovereignty over all its wealth, natural resources and economic activity,[313] sovereignty in the context of a cycle of dependence on natural resource exploitation may mean the violation of human rights in the pursuit of economic development. The political cost of Correa's pursuit of neo-extractivism has been the alienation of the indigenous support base and strong opposition to oil activity in block 43 of the Yasuní National Park, and the *XI ronda pertrolera* round of oil exploration in the southern Ecuadorian Amazon. Therefore, we have to question whether a sustainable development agenda that seeks to de-couple economic growth from development and in which all three pillars – economic, social and environmental development – are equal, is actually workable in the current neoliberal model of global governance?

## Disclosure statement

No potential conflict of interest was reported by the authors.

## Notes

1. J. McNeish and A. Borchgrevink, 'Introduction: Recovering Power from Energy – Reconsidering the Linkages Between Energy and Development', in *Contested Powers: The Politics of Energy and Development in Latin America*, ed. J. McNeish, A. Borchgrevink, and O. Locan (London: Zed Books), 2.

2. The special rapporteur takes a broad view of 'natural resources', including land, water, soil, air, coal, oil, gas, other mineral and precious metal deposits, flora and fauna, forests and timber. United Nations Human Rights Council (UNHRC), *Report of the Special Rapporteur on the Rights to Freedom of Peaceful Assembly and of Association, Maina Kiai.* A/HRC/29/25 (2015). http://www.ohchr.org/EN/HRBodies/HRC/RegularSessions/Session29/Pages/ListRep orts.aspx.

3. Ibid., 5. See also 'The Next Not-So-Cold War: As Climate Change Heats Arctic, Nations Scramble for Control and Resources', *Democracy Now*, 1 September 2015, http://www. democracynow.org/2015/9/1/the_next_not_so_cold_war.

4. UNHRC, *Report of the Special Rapporteur on the Rights to Freedom of Peaceful Assembly and of Association*, 6. See also Human Rights Watch (HRW), *At your Own Risk: Reprisals against World Bank Group Projects* (2015), https://www.hrw.org/news/2015/06/22/world-bank-group-project-critics-threatened-harassed-jailed.

5. H.D. Correa and I. Rodriguez, *Environmental Crossroads in Latin America: Between Managing and Transforming Natural Resource Conflicts* (San Jose, Costa Rica: University for Peace, 2005), 23; M. Coletta and M. Raftopoulos, eds, *Conceptualising Nature in Latin America: Multidisciplinary Approaches to Environmental Discourses* (London: ILAS, 2016).

6. Correa and Rodriguez, *Environmental Crossroads in Latin America*, 23.

7. C. Davidson and L. Kiff, 'Global Carbon-and-Conservation Models, Global Eco-States? Ecuador's Yasuní-ITT Initiative and Governance Implications', *Journal of International and Global Studies* 4, no. 2 (2013B): 1–19, 8.

8. G. O'Toole, *Environmental Politics in Latin America and the Caribbean: Introduction* (Liverpool: Liverpool University Press, 2014), 180.

9. United Nations, Economic Commission for Latin America and the Caribbean, *Conference on Sustainable Development in Latin America and the Caribbean: Follow-up to the United Nations Development Agenda Beyond 2015 and to Rio+20.* LC/L.3590/Rev.2 (2013), 7, https://sustainabledevelopment.un.org/content/documents/1001RIO_20-Rev2ing.pdf.

10. Ibid., 55.

11. O'Toole, *Environmental Politics in Latin America and the Caribbean*, 245.

12. M. Goodale and S.E. Merry, *The Practice of Human Rights Tracking Law Between the Global and the Local* (Cambridge: Cambridge University Press, 2007). Introduction available at: http://humanrights.uconn.edu/wp-content/uploads/sites/767/2014/06/MarkGoodaleHumanRi ghts.pdf, 11.

13. B. Bull and M. Aguilar-Stoen, *Environmental Politics in Latin America: Elite Dynamics, the Left Tide and Sustainable Development* (London: Routledge, 2015), 8.

14. United Nations Development Programme (UNDP), *Latin America: Energy and Sustainability* (2012), http://www.latinamerica.undp.org/content/rblac/en/home/ourwork/environmentanden ergy/overview.html.

15. A. Marin and A. Smith, *Background Paper: Towards a Framework for Analysing the Transformative Nature of Natural Resource-based Industries in Latin America: The Role of Alternatives* (research project, Sustainable Pathways for Natural Resource Industries, International Development Research Centre (IDRC), in partnership with Centro de Investigaciones para la Transformación (CENIT), 2010), 1, http://nrpathways.wix.com/home#!project-framework/c5ro.

16. A. Rebossio, 'Latinoamérica comienza a debatir la gestión de sus recursos naturales', *El País*, 5 March 2015. Available in English at: https://eyeonlatinamerica.wordpress.com/2015/03/12/ latin-america-natural-resources-management/.

17. World Bank, 'Latin America: Bridging the Gap in Water Access', *The World Bank News*, 30 August 2012, http://www.worldbank.org/en/news/feature/2012/08/30/agua-aneamiento-america-latina.

18. Rebossio, 'Latinoamérica comienza a debatir la gestión de sus recursos naturales'.

19. UNDP, *Summary: Regional Human Development Report 2013–2014* (2014), http://www. undp.org/content/undp/en/home/librarypage/hdr/human-development-report-for-latin-america -2013-2014.html.

20. Despite being one of the most unequal regions in the world, between 2002 and 2010 inequality fell in all 18 countries with the exception of Nicaragua and Costa Rica (United Nations Development Programme (UNDP), *Latin America: Poverty Reduction* (2012B), http://www.latinamerica.undp. org/content/rblac/en/home/ourwork/povertyreduction/overview.html; G.A. Cornia, 'Falling Inequality in Latin America: Policy Changes and Lessons', *The World Financial Review*, January–February (2014): 60–63).

21. The Gini index for per capita income measures the extent to which the distribution of income or consumption expenditure, among individuals or households within an economy, deviates from a perfectly equal distribution World Bank Data: *GINI index (World Bank estimate)* (2015) http://data.worldbank.org/indicator/SI.POV.GINI.

22. World Bank, *Latin American and Caribbean Overview: Results* (2015), http://www. worldbank.org/en/region/lac/overview; G. Molina, 'Inequality is Stagnating in Latin America: Should We Do Nothing?', *The Guardian*, 27 August 2014, http://www. theguardian.com/global-development-professionals-network/2014/aug/27/inequality-latin-am erica-undp.

23. Inequality plateaued in Mexico, Panama, Brazil, Dominican Republic, Chile and Paraguay between 2007 and 2012 (Molina, 'Inequality is Stagnating in Latin America').

24. J. Faieta, *UNDP: Economic Growth is Not Enough*, 20 February 2015, http://www. latinamerica.undp.org/content/rblac/en/home/ourperspective/ourperspectivearticles/2015/02/ 20/con-crecimiento-econ-mico-no-basta-jessica-faieta-.html.

25. K. Koenig, 'Ecuador Breaks Its Amazon Deal', *The New York Times*, 11 June 2014, http:// www.nytimes.com/2014/06/12/opinion/ecuador-breaks-its-amazon-deal.html?_r=0.

26. Barcena (2015), cited in Rebossio, 'Latinoamérica comienza a debatir la gestión de sus recursos naturales'.

27. International Policy Centre for Inclusive Growth (IPCIG), 'Putting National Resources Industries to Work for Sustainable Development in Latin America' (United Nations Development Programme Seminar, 18 May 2015, Brasilia), http://pressroom.ipc-undp.org/putting- national-resources-industries-to-work-for-sustainable-development-in-latin-america/.

28. Barcena (2015), cited in Rebossio, 'Latinoamérica comienza a debatir la gestión de sus recursos naturales'.

29. United Nations Agenda 2030, *Transforming our World: The 20130 Agenda for Sustainable Development*, United Nations A/RES/70/1 (2015), Target 9.4. https://sustainabledevelopme nt.un.org/post2015/transformingourworld.

30. Ibid., Declaration, para. 27.

31. O'Toole, *Environmental Politics in Latin America and the Caribbean*, 245.

32. Ibid., 188.

33. L.C. Gray and W.G. Moseley, 'A Geographical Perspective on Poverty-Environment Interactions', *The Geographical Journal* 171, no. 1 (2005): 9–23, 18.

34. United Nations Agenda 2030, *Transforming our World*.

35. Ban Ki Moon, cited in United Nations News, '"Now is the Time for Implementation", Ban Urges Session on Integrating UN Sustainability Agenda', *UN News Centre*, 2 May 2016, http://www.un.org/apps/news/story.asp?NewsID=53834#.VzinB2Nlm8U.

36. P. Martin, *Oil in the Soil: The Politics of Paying to Preserve the Amazon* (Plymouth: Rowan and Littlefield, 2011), 2.

37. A. Cori and S. Monni, *The Resource Curse Hypothesis: Evidence from Ecuador*. SEEDS Working Paper Series No 2814 (2014), 18, http://www.sustainability-seeds.org/papers/ RePec/srt/wpaper/2814.pdf.

38. U. Villalba, 'Buen Vivir vs Development: A Paradigm Shift in the Andes?', *Third World Quarterly* 34, no. 8 (2013): 1427–42, 1439.

39. L. Rival, *The Yasuní-ITT Initiative: Oil Development and Alternative Forms of Wealth Making in the Ecuadorian Amazon*. QEH Working Paper Series – QEHWPS180 (2009), 5, 4, http:// www3.qeh.ox.ac.uk/pdf/qehwp/qehwps180.pdf.

40. C. Mikkelsen et al., eds, *The Indigenous World 2014* (Copenhagen: The International Work Group for Indigenous Affairs (IWGIA), 2014), 148.

41. CODENPE (2003), Greene (2008), Radcliffe (2013), Acosta (2013), Zorrilla (2014), cited in ibid., 2.

42. J. Dayot, *Valuation Struggles in the Ecuadorean Amazon: Beyond Indigenous Peoples' Responses to Oil Extraction* (MPhil Thesis draft paper, University of Oxford, 2015).

43. A. Acosta, 'The Yasuní-ITT Initiative, or The Complex Construction of Utopia', in *The Wealth of the Commons: A World Beyond Market and State*, ed. D. Bollier and S. Hellfrich (Massachusetts: Levellers Press, 2012).

44. The first constitution in the world to do so. See also C. Kendall, 'A New Law of Nature', *The Guardian*, 24 September 2008, http://www.theguardian.com/environment/2008/sep/24/equador.conservation.

45. P. Martin and I. Scholz, 'Policy Debate | Ecuador's Yasuní-ITT Initiative: What Can We Learn from its Failure?', *International Development Policy* 5, no. 2 (2014): 3.

46. C. Larrea and L. Warnars, 'Ecuador's Yasuní-ITT Initiative: Avoiding Emissions by Keeping Petroleum Underground', *Energy for Sustainable Development* 13, no. 3 (2009): 219–23; C. Davidson and L. Kiff, 'Ecuador's Yasuní-ITT Initiative and New Green Efforts – Away from Petroleum Dependency?', *University of Calgary Occasional Papers* 3, no. 1 (2013A): 1–30, 10.

47. A. Acosta et al., 'Dejar el crudo en tierra o la búsqueda del paraíso perdido. Elementos para una propuesta política y económica para la Iniciativa de no explotación del crudo del ITT', *Revista de la Universidad Bolivariana* 8, no. 23 (2009): 429–52; Reproduced in English at: http://www.sosyasuni.org/en/index.php?option=com_content&view=article&id=130:leaving-the-oil-underground-or-the-search-for-paradise-lost&catid=17:generala, 1.

48. Ibid.

49. Martin and Scholz, 'Policy Debate | Ecuador's Yasuní-ITT Initiative', 5.

50. Ibid.

51. Through the CDM, industrialised countries invest in sustainable development projects in developing countries to earn credits which they can use to offset their greenhouse gas emission targets, set by the Kyoto Protocol: L. Warnars, *The Yasuni-ITT Initiative: An International Environmental Equity Mechanism?* (Master thesis, Political and Social Sciences of the Environment, School of Management, Radboud University Nijmegen, 2010), v, http://www.campusvirtual.uasb.edu.ec/uisa/images/yasuni/documentos/2010%20warnars%20equity.pdf.

52. For a summary of the debates surrounding REDD see also Larrea and Warnars, 'Ecuador's Yasuni-ITT Initiative', 219–23, 222.

53. Ibid.; Davidson and Kiff, 'Ecuador's Yasuní-ITT Initiative and New Green Efforts', 1–30, 11.

54. Warnars, *The Yasuni-ITT Initiative*, v. For a list of International and national NGOs who formed transnational networks to contribute to the Initiative see also Martin, *Oil in the Soil*, 2. For a full list of official backers of the Initiative see also C. Larrea, *Ecuador's Yasuní-ITT Initiative: A Critical Assessment. Environmental Governance in Latin America and the Caribbean*, (ENGOV) Project Conference, 13–16 June 2012, 12, http://www.ecuadoramazonia.com/images/documentos/Ecuadors%20Yasuni-ITT%20Initiative%20A%20Critical%20Assessment.pdf.

55. US$19 m was attributed to international donors, US$50 m was pledged as a donation by Italy and US$47 m was pledged as bilateral technical assistance by the German government. A national fundraiser in Ecuador collected about US$3 m. *El Comercio* (2012), *PRNewswire* (2012), cited in Davidson and Kiff, 'Ecuador's Yasuní-ITT Initiative and New Green Efforts ', 1–30, 25.

56. J. Watts, 'Ecuador Approves Yasuni National Park Oil Drilling in Amazon Rainforest. Environmentalists Devastated as President Blames Lack of Foreign Support for Collapse of Pioneering Conservation Plan', *The Guardian*, 16 August 2013.

57. UNDP, *UNDP Statement on Decision by Government of Ecuador to Conclude Yasuní-ITT Initiative*, 16 August 2013, http://www.undp.org/content/undp/en/home/presscenter/articles/2013/08/16/undp-statement-on-decision-by-government-of-ecuador-to-conclude-yasun-itt-initiative.html.

58. J. Vidal, 'Ecuador Spied on Amazon Oil Plan Opponents, Leaked Papers Suggest', *The Guardian*, 3 August 2015.

59. For the reasons given by the NEC for not accepting signatures see also A. Vaughan, 'Ecuador Signs Permits for Oil Drilling in Amazon's Yasuni National Park: Companies Could Start Extracting Oil Underneath Key Biodiversity Reserve on Earth by 2016', *The Guardian*, 23 May 2014.

60. Ibid.

61. Mikkelsen et al., *The Indigenous World 2014*, 153.

62. Davidson and Kiff, 'Global Carbon-and-Conservation Models, Global Eco-States?', 1–19, 11.

63. Martin, *Oil in the Soil*, 86.

64. For full data on the biodiversity, including endangered and near-threatened species see also M.S. Bass et al., 'Global Conservation Significance of Ecuador's Yasuní National Park', *PLoS ONE* 5, no. 1 (2010): 5.

65. The uniqueness of the National Park is its likelihood to maintain wet, rainforest conditions as climate change-induced drought intensifies in the eastern Amazon, increasing its ability to sustain this biodiversity in the long term. For full data on the biodiversity, including endangered and near-threatened species see ibid.

66. Deforestation within Yasuní National Park is estimated at a rate of 0.11% per year, with that rate increasing. Ibid., 7.

67. ANDES, *President Correa: Our Best Decision Was to Develop ITT Field*. Agencia Publica de Notices del Ecuador y Suramerica, 14 July 2016, http://www.andes.info.ec/en/news/president-correa-our-best-decision-was-develop-itt-field.html.

68. Swords (2014), cited in H. Guardado, 'Nicaragua's Proposed Interoceanic Canal: A Threat to the Environment and Indigenous Rights', *The Huffington Post*, 3 November 2014, http://www.huffingtonpost.com/hazel-guardado/nicaraguas-proposed-inter_b_6083274.html.

69. M. Becker, 'The Stormy Relations between Rafael Correa and Social Movements in Ecuador', *Latin American Perspectives* 190 40, no. 3 (2013): 43–62, 45.

70. P. Gready and W. Vandenhove, eds, *Human Rights and Development in the New Millennium* (London: Routledge, 2014).

71. United Nations, Economic Commission for Latin America and the Caribbean, *Conference on Sustainable Development in Latin America and the Caribbean: Follow-up to the United Nations Development Agenda Beyond 2015 and to Rio+20*. LC/L.3590/Rev.2 (2013), https://sustainabledevelopment.un.org/content/documents/1001RIO_20-Rev2ing.pdf.

72. McIber (2015), cited in UNHRC, Accountability Mechanisms for Implementing the Sustainable Development Goals: A High-Level Roundtable Discussion at the 29th Session of the UN Human Rights Council 18 June 2015, Palais des Nations, Geneva, Switzerland, http://www.fes-globalization.org/geneva/Human%20rights.htm.

73. O'Donaghue (2015), cited in ibid.

74. Ruckner (2015), cited in ibid.

75. United Nations Agenda 2030, *Transforming our World*, preamble.

76. Ban Ki Moon, cited in United Nations News, 'Civil Society Must Be "Equal Partners" in Implementing UN Sustainability Agenda, Ban Tells Parliamentarians', *UN News Centre*, 31 August 2015, http://www.un.org/apps/news/story.asp?NewsID=51759#.VfaifrQ-Cu4.

77. Office of the High Commissioner for Human Rights, *Climate Change: UN Expert Welcomes Historic Paris Agreement But Calls on States to Scale Up Efforts to Meet the 1.5 °C Target*, 21 April 2016, http://www.ohchr.org/EN/NewsEvents/Pages/DisplayNews.aspx?NewsID=19850&LangID=E.

78. Office of the High Commissioner for Human Rights, '*No Time for Complacency' – UN Rights Expert Says as the Paris Agreement Faces its First Key Test*, 13 May 2016, http://www.ohchr.org/EN/NewsEvents/Pages/DisplayNews.aspx?NewsID=19962&LangID=E.

79. Bull and Aguilar-Stoen, *Environmental Politics in Latin America*, 7.

80. B. Adams, *Green Development: Environment and Sustainability in a Developing World* (London: Routledge, 2008).

81. A. Escobar, *Encountering Development: The Making and Unmaking of the Third World* (Princeton NJ: Princeton University Press, 2011).

82. B. Mansfield, 'Sustainability', in *A Companion to Environmental Geography*, ed. N. Castree et al. (Chichester: Blackwell Publishing, 2009).

83. J. Donnelly, 'Human Rights, Democracy, and Development', *Human Rights Quarterly* 21, no. 3 (1999): 608–32, 611.

84. United Nations News, 'Global Investors Must Play Full Role in Shifting World to Clean Energy, Says UN Chief', *UN News Centre*, 27 January 2016, http://www.un.org/apps/news/story.asp?NewsID=53101#.VziuhmNlmqA.

85. United Nations News, 'World of Business Must Play its Part in Achieving New Sustainable Development Goals – UN Chief', 20 January 2016, http://www.un.org/apps/news/story.asp?NewsID=53055#.VzivBmNlk_U.

86. Ibid.

87. J. McNeish and A. Borchgrevink, 'Introduction: Recovering Power from Energy – Reconsidering the Linkages Between Energy and Development', in *Contested Powers: The Politics of Energy and Development in Latin America*, ed. J. McNeish, A. Borchgrevink, and O. Locan (London: Zed Books, 2016), 4–6.
88. N. Stammers, *Human Rights and Social Movements* (London: Pluto Press, 2009), 205.
89. R.M. Auty, *Sustaining Development in Mineral Economies: The Resource Curse Thesis* (London: Rutledge, 1993).
90. T.L. Karl, 'The Vicious Cycle of Inequality in Latin America', in *What Justice? Whose Justice? Fighting for Fairness in* Latin America, ed. S.E. Eckstein and Wickham-Crowley (Berkley, CA: University of California Press, 2003).
91. O'Toole, *Environmental Politics in Latin America and the Caribbean*, 186.
92. See also C. Brunnschweiler, 'Cursing the Blessings? Natural Resource Abundance, Institutions, and Economic Growth', *World Development* 36, no. 3 (2008): 399–419; M. Alexeev and R. Conrad, 'The Elusive Curse of Oil', *The Review of Economics and Statistics* 91, no. 3 (2009): 586–98; Mako Kuwimb, 'A Critical Study of the Resource Curse Thesis and the Experience of Papua New Guinea' (PhD thesis, James Cook University, 2010), http://researchonline.jcu.edu.au/11667/.
93. G. Bridge and T. Perreault, 'Environmental Governance', in *Companion to Environmental Geography*, ed. N. Castree et al. (Oxford: Blackwell, 2009), 475–97, 492. See also T.L. Lewis, *Ecuador's Environmental Revolutions: Ecoimperialists, Ecodependents and Ecoresisters* (Cambridge MA: The MIT Press, 2016).
94. M. Freeman, *Human Rights* (Cambridge: Polity Press, 2002), 151; G. Rist, *The History of Development: From Western Origins to Global Faith* (London: Zed Books, 2008).
95. N. Connolly, 'Corporate Social Responsibility: A Duplicitous Distraction?', *The International Journal of Human Rights* 16, no. 8 (2012): 1228–49, 1242; S. Marks and A. Clapham, *Human Rights Lexicon* (Oxford: Oxford University Press, 2005), 429.
96. Bull and Aguilar-Stoen, *Environmental Politics in Latin America*, 3.
97. O'Toole, *Environmental Politics in Latin America and the Caribbean*, 199.
98. Ibid., 245. See also A. Westervelt, 'Lawsuit Against El Salvador Mining Ban Highlights Free Trade Pitfalls', *The Guardian* 27 May 2015, https://www.theguardian.com/sustainable-business/2015/may/27/pacific-rim-lawsuit-el-salvador-mine-gold-free-trade.
99. Donella H. Meadows et al., *The Limits to Growth: A Report for the Club of Rome's Project on the Predicament of Mankind* (New York: Universe Books, 1972).
100. G. Turner, *A Comparison of the Limits to Growth with Thirty Years of Reality* (Canberra: Commonwealth Scientific and Industrial Research Organisation (CSIRO), 2008), 38.
101. D. Short et al., 2014. 'Extreme Energy, "Fracking" and Human Rights: A New Field for Human Rights Impact Assessments?', *The International Journal of Human Rights Special Issue: Corporate Power and Human Rights* (December 2014), 2.
102. Latin America is not considered a main contributor to greenhouse gasses having contributed only about 4% of global emissions. O'Toole, *Environmental Politics in Latin America and the Caribbean*, 154–5.
103. Ibid.
104. Ibid., 180.
105. Davidson and Kiff, 'Global Carbon-and-Conservation Models, Global Eco-States?', 1–19, 5.
106. J. Lessmann et al., 'Large Expansion of Oil Industry in the Ecuadorian Amazon: Biodiversity Vulnerability and Conservation Alternatives', *Ecology and Evolution* 6, no. 14 (2016): 4997–5012.
107. Larrea and Warnars, 'Ecuador's Yasuní-ITT Initiative'; Davidson and Kiff, 'Ecuador's Yasuní-ITT Initiative and New Green Efforts', 1–30, 6.
108. Daily crude oil production is measured in barrels per day. Since 1970 Ecuador has increased its crude oil production per day by 12,183%, versus a 44% increase from OPEC overall. Lewis, *Ecuador's Environmental Revolutions*, 31.
109. R. Ray and A. Chimienti, *A Line in the Equatorial Forests: Chinese Investment and the Environmental and Social Impacts of Extractive Industries in Ecuador*. Discussion Paper 2015–16. Working Group on Development and Environment in the Americas. Global Economic Governance Initiative (ENGOV), 2014, 8, http://www.bu.edu/pardeeschool/files/2014/12/Ecuador1.pdf.

110. Economic Commission for Latin America and the Caribbean (ECLAC), *Economic Survey of Latin America, 2012: Policies for an Adverse International Economy* (Washington, DC: United Nations Publications, 2012), http://www.cepal.org/en/node/30073.

111. Neo-extractivist governments are characterised by the expansion of industries for an export-led development model to fund social programmes through higher taxes or nationalisation. J.R. Webber, 'Neostructuralism, Neoliberalism, and Latin America's Resurgent Left', *Historical Materialism: Research in Critical Marxist Theory* 18, no. 3 (2010): 208–29.

112. L. Shade, 'Sustainable Development or Sacrifice Zone? Politics Below the Surface in Post-neoliberal Ecuador', *The Extractive Industries and Society* 2, no. 4 (2015): 775–85, 773.

113. World Bank, *Ecuador Overview: Context* (Washington, DC: World Bank, 2015), http://www.worldbank.org/en/country/ecuador/overview.

114. Monaghan (2008), cited in A. Kennemore and G. Weeks, 'The Elusive Search for a Post-Neo-liberal Development Model in Bolivia and Ecuador', *Bulletin of Latin American Research* 30, no. 3 (2011): 267–81, 276.

115. World Bank, *Ecuador Overview: Context*.

116. E. Sinnott, J. Nash, and A. de la Torre, *Natural Resources in Latin America and the Caribbean: Beyond Booms and Busts?* (Washington, DC: World Bank Publications, 2010).

117. B. Bull and M. Aguilar-Stoen, *Environmental Politics in Latin America: Elite Dynamics, the Left Tide and Sustainable Development* (London: Rutledge, 2015), 4.

118. Correa (2009), quoted in *El Cuidadano* in Kennemore and Weeks, 'The Elusive Search for a Post-Neoliberal Development Model in Bolivia and Ecuador', 267–81, 274.

119. Republic of Ecuador, *Constitution of the Republic of Ecuador*. Political Database of the Americas (2008), 279, http://pdba.georgetown.edu/Constitutions/Ecuador/english08.htmlart.

120. Ibid.

121. C. Walsh, 'Development as Buen Vivir: Institutional Arrangements and (de)colonial Entanglements', *Development* 53, no. 1 (2010): 15–21, 18.

122. Republic of Ecuador, *Constitution of the Republic of Ecuador*, 275.

123. Gudynas (2011), cited in D. de Wit, 'UN-REDD and the Yasuní-ITT Initiative as Global Environmental Governance Mechanisms', *Bridges* 7 (Spring 2013), 3–4, https://www.coastal.edu/media/academics/bridges/pdf/deWitFINAL.pdf.

124. M. Cunningham, '"Living Well" The Latin American Perspective', *Indigenous Affairs* (2010) 1–2/10, 52–9, 53.

125. Hevia-Pacheco and Vergara-Camus (2013), cited in A. Ordóñez et al., *Sharing the Fruits of Progress: Poverty Reduction in Ecuador* (London: Overseas Development Institute, 2015), 29, https://www.odi.org/publications/9958-ecuador-extreme-poverty-progress-reduction-inequality.

126. National Secretariat of Planning and Development (SENPLADES), *National Development Plan/National Plan for Good Living, 2013–2017* (Quito, Ecuador: 2013), 460–1.

127. Ibid.

128. C. Larrea, *Ecuador's Yasuni-ITT Initiative: A Critical Assessment. Environmental Governance in Latin America and the Caribbean*, (ENGOV) Project Conference, 13–16 June 2012, 12–13, http://www.ecuadoramazonia.com/images/documentos/Ecuadors%20Yasuni-ITT%20Initiative%20A%20Critical%20Assessment.pdf.

129. A. Acosta et al., 'Dejar el crudo en tierra o la búsqueda del paraíso perdido. Elementos para una propuesta política y económica para la Iniciativa de no explotación del crudo del ITT', *Revista de la Universidad Bolivariana* 8, no. 23 (2009): 429–52. Reproduced in English at: http://www.sosyasuni.org/en/index.php?option=com_content&view=article&id=130:leaving-the-oil-underground-or-the-search-for-paradise-lost&catid=17:generala, 5.

130. V. Davidov, 'Saving Nature or Performing Sovereignty? Ecuador's Initiative to "Keep Oil in the Ground"', *Anthropology Today* 28, no. 3 (2012): 12–16, 12.

131. Walsh, 'Development as Buen Vivir', 15–21; L. Rival, 'Ecuador's Yasuni ITT Initiative: The Old and New Styles of Petroleum', *Ecological Economics* 70 (2010): 358–65.

132. U. Villalba, 'Buen Vivir vs Development: A Paradigm Shift in the Andes?', *Third World Quarterly* 34, no. 8 (2013): 1427–42, 1439.

133. L. Rival, *The Yasuní-ITT Initiative: Oil Development and Alternative Forms of Wealth Making in the Ecuadorian Amazon*. QEH Working Paper Series – QEHWPS180 (2009), 5, http://www3.qeh.ox.ac.uk/pdf/qehwp/qehwps180.pdf.

134. Kennemore and Weeks, 'The Elusive Search for a Post-Neoliberal Development Model in Bolivia and Ecuador', 267–81, 280.
135. Becker, 'The Stormy Relations between Rafael Correa and Social Movements in Ecuador', 43–62, 51.
136. B. Bull and M. Aguilar-Stoen, *Environmental Politics in Latin America: Elite Dynamics, the Left Tide and Sustainable Development* (London: Rutledge, 2015), 4.
137. R. Ray and A. Chimienti, *A Line in the Equatorial Forests: Chinese Investment and the Environmental and Social Impacts of Extractive Industries in Ecuador* (Boston, MA: BU Global Economic Governance Initiative Working Paper 15–17, 2014), 35, http://www.bu.edu/pardeeschool/files/2014/12/Ecuador1.pdf.
138. Ibid., 35.
139. Secretaria de Hidrocarburos del Ecuador (2011), cited in Lessmann et al., 'Large Expansion of Oil Industry in the Ecuadorian Amazon: Biodiversity Vulnerability and Conservation Alternatives', *Ecology and Evolution* 6, no. 14 (2016): 4997–5012.
140. ANDES, 'Ecuador and Chinese Consortium Andes Petroleum Sign Two Exploration Contracts in the Amazon', *ANDES*, 26 January 2016, http://www.andes.info.ec/en/news/ecuador-and-chinese-consortium-andes-petroleum-sign-two-exploration-contracts-amazon.html.
141. El Comercio (2013), cited in Lessmann et al., 'Large Expansion of Oil Industry in the Ecuadorian Amazon', 5013.
142. I. Riofrio, 'Oil Extraction Threatens to Expand Further into Ecuadorean Rainforest Under New 20-Year Contract', *Mongabay News*, 3 February 2016, https://news.mongabay.com/2016/02/oil-extraction-threatens-to-expand-further-into-ecuadorean-rainforest-under-new-20-year-contract/.
143. ANDES, 'Ecuador and Chinese Consortium Andes Petroleum Sign Two Exploration Contracts in the Amazon'.
144. J. Kaiman, 'Controversial Ecuador Oil Deal Lets China Stake an $80-Million Claim to Pristine Amazon Rainforest', *LA Times*, 29 January 2016, http://www.latimes.com/world/mexico-americas/la-fg-ecuador-china-oil-20160129-story.html.
145. ANDES, 'Ecuador and Chinese Consortium Andes Petroleum Sign Two Exploration Contracts in the Amazon'.
146. Kaiman, 'Controversial Ecuador Oil Deal Lets China Stake an $80-Million Claim to Pristine Amazon Rainforest'.
147. Ray and Chimienti, *A Line in the Equatorial Forests*, 1.
148. C. Mikkelsen et al., eds., *The Indigenous World 2014* (Copenhagen: The International Work Group for Indigenous Affairs (IWGIA), 2014), 150.
149. M. Arsel and N.A. Angel, '"Stating" Nature's Role in Ecuadorian Development: Civil Society and the Yasuní-ITT Initiative', *Journal of Developing Societies* 28, no. 2 (2012): 203–27, 207.
150. F.E. Lu and N.L. Silva, 'Imagined Borders: (Un)Bounded Spaces of Oil Extraction and Indigenous Sociality in "Post-Neoliberal" Ecuador', *Social Sciences* 4, no. 2 (2015): 434–58, 435.
151. Becker, 'The Stormy Relations between Rafael Correa and Social Movements in Ecuador', 54.
152. A. Ordóñez, E. Samman, C. Mariotti, and I. Marcelo Borja, *Sharing the Fruits of Progress: Poverty Reduction in Ecuador* (London: Overseas Development Institute, 2015), 30, https://www.odi.org/publications/9958-ecuador-extreme-poverty-progress-reduction-inequality.
153. S. Nicolai, T. Bhatkal, C. Hoy, and T. Aedy, *Projecting Progress: The SDGs in Latin America and the Caribbean Regional Scorecard* (London: Overseas Development Institute June 2016, 2016), 19, https://www.odi.org/publications/10454-projecting-progress-sdgs-latin-america-and-caribbean.
154. CEPAL, cited in Ordóñez et al., *Sharing the Fruits of Progress*, 32.
155. Ray and Chimienti, *A Line in the Equatorial Forests*, 6.
156. Asemblea Nacional del Ecuador (2010b) art. 94, cited in Ray and Chimienti, *A Line in the Equatorial Forests*, 7.
157. The Palma ratio is the ratio of the average income of the richest 10% in a country compared to the average income of the poorest 40%. Cobham and Sumner (2013), cited in Ordóñez et al., *Sharing the Fruits of Progress*, 9.
158. The NBI combines deprivation in housing conditions, access to water and sanitation, the household dependency ratio and children's access to primary education. A household is considered poor if it is deprived in one or more of these dimensions. INEC (2015a), cited in Ordóñez et al., *Sharing the Fruits of Progress*, 16.

159. Ray and Chimienti, *A Line in the Equatorial Forests*, 7.
160. Becker, 'The Stormy Relations between Rafael Correa and Social Movements in Ecuador', 54
161. Ray and Chimienti, *A Line in the Equatorial Forests*, 1.
162. The financial crisis that began in 2008 put pressure on Ecuador's international sources of finance. In 2008, Ecuador defaulted on two outstanding bonds totalling $3.2 billion dollars leading Moody's to downgrade Ecuador's debt to Caa3, and Ecuador losing access to its traditional Western creditors. Therefore, China has seen Ecuador through a prolonged period of limited access to financial markets. Porzecanski (2010), cited in Ray and Chimienti, *A Line in the Equatorial Forests*, 12.
163. P. Martin and I. Scholz, 'Policy Debate | Ecuador's Yasuní-ITT Initiative: What Can We Learn from its Failure?', *International Development Policy* 5, no. 2 (2014): 4. In an interview with the foreign media on 16 February 2014, President Correa said there was no limit to the loans from China, 'the more they can lend us, the better. We need financing for development and we have profitable projects. (…) We complement China, they have a surplus of liquidity and a shortage of hydrocarbons while we have a surplus of hydrocarbons and a shortage of liquidity. China finances the USA and could pull Ecuador out of underdevelopment', Correa (2014), cited in A. Cori and S. Monni, 'The Resource Curse Hypothesis: Evidence from Ecuador', SEEDS Working Paper Series No 2814 (2014), 12, http://www.sustainability-seeds.org/papers/RePec/srt/wpaper/2814.pdf.
164. Reuters (2013), cited in Cori and Monni, 'The Resource Curse Hypothesis', 11.
165. The official document, in its entirety, can be consulted at the following link: http://www.theguardian.com/environment/interactive/2014/feb/19/china-development-bank-credit-proposal-oil-drilling-ecuador1. Cori and Monni, 'The Resource Curse Hypothesis', 11.
166. Normally, countries that are part of OPEC, like Ecuador, do not offer this possibility for competitive reasons. According to Reuters (2013), EP PetroEcuador warned the government in March 2011 that PetroChina's claim to Ecuador's oil supply could prevent the country from signing more favourable contracts. Cori and Monni, 'The Resource Curse Hypothesis', 11.
167. The credit line was part of a broader package, including China's Ex-Im Bank extending a $5.3bn credit line to help maintain public spending, and another $1.5 bn loan from China Development Bank, that Correa secured on a visit to Beijing in 2015. 2016 data from the finance ministry, which excludes the latest ICBC credit line, showed that Beijing was owed $5.4bn by Ecuador and that a 'chunk' of the China loans are backed by oil exports. M. Badkar, 'Ecuador Secures $970m Credit Line From China's ICBC', *Financial Times*, 22 January 2016, https://www.ft.com/content/87d39bc8-4bad-3d14-9f57-50c083b3c916.
168. A. Zuckerman, (2016), cited in I. Riofrio, 'Oil Extraction Threatens to Expand Further into Ecuadorean Rainforest under New 20-Year Contract', *Mongabay News*, 3 February 2016, https://news.mongabay.com/2016/02/oil-extraction-threatens-to-expand-further-into-ecuadorean-rainforest-under-new-20-year-contract.
169. GFC Media Group, 'China's Interest in Latin America Grows Exponentially', 16 February 2016, http://www.gfcmediagroup.com/news/article/418/chinas-interest-in-latin-america-grows-expone. While analysts agree that by and large LAC nations have to pay a higher premium for loans from China, on the whole that higher premium is in the form of interest rates, not loans-for-oil. The majority of Chinese loans-for-oil in Latin America are linked to market prices, not quantities of oil. K. Gallagher, A. Irwin, and K. Koleski, *The New Banks in Town: Chinese Finance in Latin America*, Inter-American Dialogue Report, February 2012, http://ase.tufts.edu/gdae/Pubs/rp/GallagherChineseFinanceLatinAmerica.pdf.
170. Larrea et al. (2010), cited in Lessmann et al., 'Large Expansion of Oil Industry in the Ecuadorian Amazon', 5012.
171. Ray and Chimienti, *A Line in the Equatorial Forests*, 8–9. China plays only a minor role in Ecuador's export market currently, buying just 3.5% of Ecuador's exports from 2008 to 2012, however its role in international trade and investment in Latin American and Ecuador is growing in importance, and as it does so, it is increasing petroleum's importance in Ecuador's overall export basket. Ray and Chimienti, 10.
172. T.L. Karl, 'The Perils of the Petro-State: Reflections on the Paradox of Plenty', *International Affairs* 53 (1999): 31–48, 36.
173. R. Ray, 'China in Latin America Seeking a Path Toward Sustainable Development', *ReVista* (Fall 2015): 20–2.

174. T.L. Karl, 'The Vicious Cycle of Inequality in Latin America', in *What Justice? Whose Justice? Fighting for Fairness in* Latin America, ed. S. E. Eckstein and T. P. Wickham-Crowley (Berkley: University of California Press, 2003).

175. V. Davidov, 'Saving Nature or Performing Sovereignty? Ecuador's Initiative to "Keep Oil in the Ground"', *Anthropology Today* 28, no. 3 (2012): 12–16, 12.

176. The 'Dutch disease' phenomenon identifies nations that primarily export raw commodities as having overvalued currencies because their exports' prices are determined by the world market rather than by manufacturing costs. This issue of overvaluing currency is compounded in Ecuador by its use of the US dollar as its national currency. Ray and Chimienti, *A Line in the Equatorial Forests*, 7.

177. National Secretariat of Planning and Development (SENPLADES), *National Development Plan/National Plan for Good Living, 2013–2017* (Quito, Ecuador, 2013), 331.

178. E. Sinnott, J. Nash, and A. de la Torre, *Natural Resources in Latin America and the Caribbean: Beyond Booms and Busts?* (Washington, DC: World Bank Publications, 2010), 37.

179. Ray and Chimienti, *A Line in the Equatorial Forests*, 7.

180. F. Dawson, 'Ecuador U-turns on Nationalisation', *Pathfinder Buzz*, 18 April 2013, http://pathfinderbuzz.com/ecuador-u-turns-on-nationalisa-tion/. Lack of economic diversification also affects the energy supply. Ecuador has a vast hydroelectric potential, mostly in the Andean mountains, estimated at 21,122 MW, and also has the potential for other renewable energy sources, mostly geothermal; solar and wind power are also regarded as important. Larrea and Warnars, 'Ecuador's Yasuní-ITT Initiative', 220.

181. Becker, 'The Stormy Relations between Rafael Correa and Social Movements in Ecuador', 54.

182. Ordóñez et al., *Sharing the Fruits of Progress*, 29.

183. Kimerling (1991), Rosenfeld et al. (1997), Fontaine (2003), San Sebastian and Hurtig (2004), Bravo (2007), Finer et al. (2008), De la Bastida (2009), and Larrea et al. (2010), cited in Lessmann et al., 'Large Expansion of Oil Industry in the Ecuadorian Amazon', 5002.

184. Kimerling (1991) and Domınguez (2010), cited in Lessmann et al., 'Large Expansion of Oil Industry in the Ecuadorian Amazon', 5002.

185. Fontaine (2003); Sierra (2000), and Bilsborrow et al. (2004), cited in Lessmann et al., 'Large Expansion of Oil Industry in the Ecuadorian Amazon', 5002.

186. O'Rourke and Connolly (2003), Colectivo de Geografia Critica de Ecuador (2014), cited in Lessmann et al., 'Large Expansion of Oil Industry in the Ecuadorian Amazon', 5002.

187. Suarez et al. (2009), Ponce-Reyes et al. (2013), and Espinosa et al. (2014), cited in Lessmann et al., 'Large Expansion of Oil Industry in the Ecuadorian Amazon', 5002.

188. Lessmann et al., 'Large Expansion of Oil Industry in the Ecuadorian Amazon', 5002.

189. Republic of Ecuador, *Constitution of the Republic of Ecuador* (Political Database of the Americas, 2008), art. 71, http://pdba.georgetown.edu/Constitutions/Ecuador/english08.html.

190. Ibid., art. 313.

191. Ibid., art. 407.

192. Ibid.

193. Lessmann et al., 'Large Expansion of Oil Industry in the Ecuadorian Amazon', 5006.

194. Ibid., 5012.

195. Ibid., 5003.

196. Kimerling (1991), Rosenfeld et al. (1997), Fontaine (2003), San Sebastian and Hurtig (2004), and Finer et al. (2008), cited in ibid., 5005.

197. Bass et al. 'Global Conservation Significance of Ecuador's Yasuní National Park'; C. Larrea, A.I. Larrea, A.L. Bravo, P. Belmont, C. Baroja, and C. Mendoza et al. *Petroleo, sustentabilidad y desarrollo en la Amazon ıa Centro-Sur. Fundacio n Pachamama* (Quito, Ecuador: Unidad de Informacio n Socio Ambiental de la Universidad Andina Simon Bolıvar, 2010); CONFENIAE and CONAIE (2012), cited in Lessmann et al., 'Large Expansion of Oil Industry in the Ecuadorian Amazon', 5003.

198. Lessmann et al., 'Large Expansion of Oil Industry in the Ecuadorian Amazon', 5012.

199. Ibid.

200. Ray and Chimienti, *A Line in the Equatorial Forests*, 19–20.

201. Oil pollution in Ecuador has been characterised as 'one of the largest environmental disasters in history' by Rainforest Action Network, Business & Human Rights Resource Centre, *Human Rights Impacts of Oil Pollution: Ecuador Impacts on Health, Livelihoods, Environment* (2016), https://business-humanrights.org/en/human-rights-impacts-of-oil-pollution-ecuador-22.

202. Today the Ecuadorian Amazon is suffering a public health crisis of immense proportions. The root cause of this crisis is water contamination from 40 years of oil operations. The oil infrastructure developed and operated by Texaco had utterly inadequate environmental controls, and consequently Texaco dumped 18 billion gallons of toxic wastewater directly into the region's rivers. The contamination of water essential for the daily activities of thousands of people has resulted in an epidemic of cancer, miscarriages, birth defects and other ailments. Chevron Toxico, *Chevron Toxico – The Campaign for Justice in Ecuador: Health Impacts* (2016), http://chevrontoxico.com/about/health-impacts/. See also Business & Human Rights Resource Centre, *Texaco/Chevron Lawsuits (re Ecuador)* (2016), https://business-humanrights.org/en/texacochevron-lawsuits-re-ecuador.

203. R. Ray, P. Gallagher, A. Lopez, and C. Sanborn, *China in Latin America: Lessons or Spout-South Cooperation and Development* (Boston University Global Economic Governance Initiative, 2015), 13, http://www.bu.edu/pardeeschool/files/2014/12/Working-Group-Final-Report.pdf.

204. In the extractive industry Greenfield projects refer to new projects that include the exploration of previously unexplored areas, or in areas where oil, gas or mineral deposits are not already known to exist. These projects rely on the predictive power of oil/gas/ore genesis models. Schlumberger, *Oilfield Glossary* (2016), http://www.glossary.oilfield.slb.com/en/Terms/g/greenfield.aspx; 'Mineral Exploration Companies – Greenfield Exploration vs. Brownfield Exploration', *undervaluedequity.com* (2015), http://www.undervaluedequity.com/Mineral-Exploration-Companies-Greenfield-Exploration-vs.-Brownfield-Exploration.html.

205. Ray and Chimienti, *A Line in the Equatorial Forests*, 27.

206. Ibid., 24.

207. Bass et al., 'Global Conservation Significance of Ecuador's Yasuní National Park', 13.

208. Repsol (2012), cited in Ray and Chimienti, *A Line in the Equatorial Forests*, 28.

209. D. Hill, 'Ecuador: Oil Company Has Built "Secret" Road Deep into Yasuni National Park', *The Ecologist*, 6 June 2014, http://www.thee-cologist.org/News/news_analysis/2426486/ecuador_oil_company_has_built_secret_road_deep_into_yasuni_national_park.html.

210. Bass et al., 'Global Conservation Significance of Ecuador's Yasuní National Park', 7.

211. Greenberg et al. (2005), cited in Ray, and Chimienti, *A Line in the Equatorial Forests*, 28.

212. Blue Moon Fund, *Offshore-Inland: Advanced Oil and Gas Technology to Save Tropical Forests and Indigenous Cultures from Destruction*, http://www.bluemoonfund.org/wp-content/uploads/2015/09/offshore-inland-english.pdf. See also J. Tollefson, 'Fighting for the Forest: The Roadless Warrior', *Nature* 480 (2011): 22–4, http://www.nature.com/news/fighting-for-the-forest-the-roadless-warrior-1.9494. See also http://www.bluemoonfund.org/projects/saving-the-amazon-rainforest-using-modern-technology/.

213. Zuckerman (2014), cited in Ray and Chimienti, *A Line in the Equatorial Forests*, 24.

214. Bass et al. (2010), cited in ibid.

215. Ray and Chimienti, *A Line in the Equatorial Forests*, 24.

216. Espinosa (2014) and Swing, (2013), cited in ibid.

217. Auquilla (2014), cited in ibid.

218. Ray and Chimienti, *A Line in the Equatorial Forests*, 13.

219. The Chinese Ministry of Commerce (MOFCOM) has published voluntary 'Guidelines for Environmental Protection in Foreign Investment and Cooperation' for all investors, regardless of whether they are public or private, or how they are financed. While these are not binding, they carry moral authority for state-owned enterprises. For projects that are bank-financed, the China Banking Regulatory Commission (CBRC) has set 'Green Credit Guidelines' for all Chinese banks that finance investment projects abroad, which include requiring investments to meet host-country and international environmental laws. Ray et al., *China in Latin America*, 14.

220. Ray et al., *China in Latin America*, 3.

221. K. Gallagher, A. Irwin, and K. Koleski, *The New Banks in Town: Chinese Finance in Latin America*. Inter-American Dialogue Report, February 2012, http://ase.tufts.edu/gdae/Pubs/rp/GallagherChineseFinanceLatinAmerica.pdf.

222. Ray, 'China in Latin America Seeking a Path Toward Sustainable Development'.

223. See also Ray and Chimienti, *A Line in the Equatorial Forests*, 25–6.

224. L. Shade, 'Sustainable Development or Sacrifice Zone? Politics Below the Surface in Post-Neoliberal Ecuador', *The Extractive Industries and Society* 2, no. 4 (2015): 775–85, 781.

225. M. Cunningham, '"Living Well" The Latin American Perspective', *Indigenous Affairs* 1–2/10 (2010): 52–9, 53.
226. Larrea, *Ecuador's Yasuni-ITT Initiative*, 12.
227. Cunningham, '"Living Well" The Latin American Perspective', 53.
228. Republic of Ecuador, *Constitution of the Republic of Ecuador*, art. 57. According to the 2010 census, indigenous people represent 7% of the Ecuadorian population (1.018 million), comprising 14 distinct peoples and 12 indigenous cultures (Minority Rights Group, *Ecuador Overview*, http://www.mi-norityrights.org/4133/ecuador/ecuador-overview.html; Larrea and Warnars, 'Ecuador's Yasuni-ITT Initiative', 220).
229. United Nations Declaration on the Rights of Indigenous Peoples (UNDRIPS), UN Doc: 61/295 (2007). See also Inter-American Commission on Human Rights (IACHR), *Indigenous and Tribal Peoples' Rights over their Ancestral Lands and Natural Resources: Norms and Jurisprudence of the Inter-American Human Rights System*, Organisation of American States (OAS) OEA/Ser.L/V/II. Doc. 56/09 (2009), http://www.oas.org/en/iachr/indigenous/docs/pdf/ancestrallands.pdf.
230. United Nations Human Rights Council (UNHRC), Follow Up Report on Indigenous Peoples and the Right to Participate in Decision Making, With a Focus on Extractive Industries. A/HRC/ 21/55 (2012), para. 13, http://www.ohchr.org/Documents/HRBodies/HRCouncil/RegularSession/Session21/A-HRC-21-55_en.pdf.
231. Ray and Chimienti, *A Line in the Equatorial Forests*, 32.
232. Ibid.
233. Zuckerman (2016), cited in I. Riofrio, 'Oil Extraction Threatens to Expand Further into Ecuadorean Rainforest under New 20-Year Contract', *Mongabay News*, 3 February 2016, https://news.mongabay.com/2016/02/oil-extraction-threatens-to-expand-further-into-ecuadorean-rainforest-under-new-20-year-contract/.
234. Melo (2014), cited in Ray and Chimienti, *A Line in the Equatorial Forests*, 33.
235. Christopher Moseley, ed., *Atlas of the World's Languages in Danger*, 3rd edn (Paris: UNESCO Publishing, 2010).
236. UNESCO (United Nations Educational, Scientific, and Cultural Organization), 'Oral Heritage and Cultural Manifestations of the Zápara People', *Intangible Cultural Heritage* (2008), http://www.unesco.org/culture/ich/en/RL/00007.
237. Riofrio, 'Oil Extraction Threatens to Expand Further into Ecuadorean Rainforest under New 20-Year Contract'.
238. *The Mayagna (Sumo) Awas Tingni Community* v. *Nicaragua* (Judgment of 31 August 2001), (Ser. C) No. 79. Inter-American Court of Human Rights [*Awas Tingni v. Nicaragua*].
239. Organisation of American States, American Convention on Human Rights (ACHR), 'Pact of San Jose, Costa Rica' (1969), art. 21.
240. IACHR, *Indigenous and Tribal Peoples' Rights over their Ancestral Lands and Natural Resources: Norms and Jurisprudence of the Inter- American Human Rights System*. Organisation of American States (OAS) OEA/Ser.L/V/II. Doc. 56/09 (2009), para. 2, http://www.oas.org/en/iachr/indigenous/docs/pdf/ancestrallands.pdf.
241. Kaiman, 'Controversial Ecuador Oil Deal Lets China Stake an $80-Million Claim to Pristine Amazon Rainforest'.
242. Republic of Ecuador, *Constitution of the Republic of Ecuador*, art. 66.
243. Ray, 'China in Latin America Seeking a Path Toward Sustainable Development'.
244. ANDES, 'Ecuador and Chinese Consortium Andes Petroleum Sign Two Exploration Contracts in the Amazon'.
245. Ray et al., *China in Latin America*, 11.
246. SHE (2012), cited in Ray and Chimienti, *A Line in the Equatorial Forests*, 30.
247. Ray et al., *China in Latin America*, 11.
248. Ray and Chimienti, *A Line in the Equatorial Forests*, 31.
249. See also Republic of Ecuador, *Citizens Participation Law* (2010), art. 83.
250. International Labour Organisation (ILO), *C169 – Indigenous and Tribal Peoples Convention*, 1989 (No. 169), http://www.ilo.org/dyn/normlex/en/f?p=NORMLEXPUB:12100:0::NO::P12100_INSTRUMENT_ID,P12100_LANG_CODE:312314,en.
251. IACHR, Case of the *Saramaka* v. *Suriname*, Judgment of August 12, 2008 (Interpretation of the Judgement on Preliminary Objections, Merits and Reparations, and Costs (2008), para. 17, http://www.corteidh.or.cr/docs/casos/articulos/seriec_185_ing.pdf.

252. *Saramaka People* v. *Suriname.* Inter-American Court of Human Rights, Preliminary Objections, Merits, Reparations, and Costs, Judgment of 28 November 2007, Series C No. 172 [*Saramaka*]. http://www.corteidh.or.cr/docs/casos/articulos/seriec_172_ing.pdf para.137

253. U. Khatri, 'Indigenous Peoples' Right to Free, Prior, and Informed Consent in the Context of State-Sponsored Development: The New Standard Set by Sarayaku v. Ecuador and its Potential to Delegitimize the Belo Monte Dam', *American University International Law Review* 29, no. 1 (2013): 165–207, 184.

254. United Nations, United Nations Declaration on the Rights of Indigenous Peoples, art. 32, para. 2.

255. Mikkelsen et al., *The Indigenous World 2014*, 154.

256. ANDES, 'Ecuador and Chinese Consortium Andes Petroleum Sign Two Exploration Contracts in the Amazon'.

257. Melo (2014), cited in Ray and Chimienti, *A Line in the Equatorial Forests*, 33.

258. Ray and Chimienti, *A Line in the Equatorial Forests*, 27.

259. In November 2006, 300 local residents entered, occupied and stopped production for Andes Petroleum, demanding 400 local jobs. In July 2007 community members, transit workers and municipal staff from the town of Nueva Loja blocked a major road to demand more local jobs and other local investment. Andes Petroleum and PetroOriental also faced a series of lawsuits that forced them to make additional payments to a total of 307 former contract workers after original profits were shared among too few workers. As a result, the companies had to pay an additional $16 million to the originally excluded workers. Ray and Chimienti, *A Line in the Equatorial Forests*, 22.

260. Ibid., 23.

261. Extraction supports just one direct job and 16 indirect jobs per million dollars of output, compared to 25 direct and 22 indirect jobs for manufacturing. Overall, exports to China support just 30 jobs per million dollars, compared to 70 jobs per million dollars supported by Ecuadoran exports overall. Ray, 'China in Latin America Seeking a Path Toward Sustainable Development'.

262. Ibid.

263. Silvestre (2014) and Obando (2014), cited in Ray, and Chimienti, *A Line in the Equatorial Forests*, 23.

264. Ray and Chimienti, *A Line in the Equatorial Forests*, 23.

265. Radcliffe (2012), cited in Ordóñez et al., *Sharing the Fruits of Progress*, 28.

266. Guatemala (2010), cited in Arsel and Angel, '"Stating" Nature's Role in Ecuadorian Development', 218.

267. A. Keyman, 'Evaluating Ecuador's Decision to Abandon the Yasuni-ITT Initiative' (Thesis, 22 February 2015, E International Relations Students Website), 9; Mikkelsen et al., *The Indigenous World 2014*, 150.

268. C. Lang, 'Ecuador Plans to Drill for Oil in the Yasuní National Park', *REDD Monitor*, 22 August 2013, http://www.redd-monitor.org/2013/08/22/ecuador-plans-to-drill-for-oil-in-the-yasuni-national-park/.

269. UNHRC, *Report of the Special Rapporteur on the Rights to Freedom of Peaceful Assembly and of Association, Maina Kiai*, para. 11.

270. Keyman, 'Evaluating Ecuador's Decision to Abandon the Yasuni-ITT Initiative'.

271. UNHRC, *Report of the Special Rapporteur on the Rights to Freedom of Peaceful Assembly and of Association, Maina Kiai*, para. 40.

272. Ibid., para. 39.

273. Ibid., para. 11.

274. Keyman, 'Evaluating Ecuador's Decision to Abandon the Yasuni-ITT Initiative'. See also K. Koenig, 'Indigenous March Descends on Quito, as National Strike Presses for Major Reforms', *Amazon Watch*, 13 August 2015, http://amazonwatch.org/news/2015/0813-indigenous-march-descends-on-quito-as-national-strike-presses-for-major-reforms.

275. Cholango (2012), cited in Becker, 'The Stormy Relations between Rafael Correa and Social Movements in Ecuador', 43.

276. J. Watts and D. Collyns, 'Ecuador Indigenous Leader Found Dead Days Before Planned Lima Protest: Shuar Leader José Isidro Tendetza Antún Missing since 28 November. Activists Believe Death Linked to Opposition to State-Chinese Mine Project', *The Guardian*, 6 December 2014.

277. Mikkelsen et al., *The Indigenous World 2014*, 155.

278. UNHRC, *Report of the Special Rapporteur on the Rights to Freedom of Peaceful Assembly and of Association, Maina Kiai*, para. 41.
279. Ibid., para. 34.
280. Decline of open debate was of particular concern in Latin America during 2015, according to *The World Press Freedom Index*. The report highlighted 'institutional violence' in Venezuela and Ecuador, organised crime in Honduras, impunity in Colombia, corruption in Brazil and media concentration in Argentina as the main obstacles to press freedom. *The Guardian*, 'Era of Propaganda: Press Freedom in Decline, says Reporters Without Borders', 20 April 2016, http://www.theguardian.com/media/2016/apr/20/era-of-propaganda-press-freedom-in-decline-says-reporters-without-borders. See also HRW, *World Report 2015: Ecuador – Events of 2014* (2015), https://www.hrw.org/world-report/2015/country-chapters/ecuador; HRW, 'Ecuador: Courts Stalling on Protester Appeals. Apply New Rules to Overturn Groundless Convictions', *HRW News*, 21 July 2015, https://www.hrw.org/news/2015/07/21/ecuador-courts-stalling-protester-appeals.
281. Martinez (2012), cited in U. Brand, 'Energy Policy and Resource Extractivism: Resistances and Alternatives', in *Energy Policy and Resource Extractivism: Resistances and Alternatives. Seminar Reader, 24–26 March 2013*, ed. M. Gensler and M. Rosa (Brussels: Luxemburg Foundation, 2013), 6, http://www.rosalux.eu/fileadmin/user_upload/reader-en-extractivism-tunis2013.pdf.
282. Watts and Collyns, 'Ecuador Indigenous Leader Found Dead Days before Planned Lima Protest'.
283. M. Kai, 'Expert Essay: The Revolution May be Televised but Most Protests Aren't', in *ISHR Annual Report 2015* (Geneva: International Service for Human Rights (ISHR), 2014), 30, http://www.ishr.ch/sites/default/files/article/files/c_-_2015_ishr_annualreport_web.pdf.
284. Universal Rights Group (URG), *Report of the First Regional Consultation with Environmental Human Rights Defenders (EHRDs): African and European Regions* (2014), para.11, http://www.universal-rights.org/events-detail/regional-consultation-with-environmental-human-rights-defenders/.
285. Ibid.
286. A. Perez, 'Nicaragua is Not for Sale', *Cetri: Southern Social Movements Newswire*, 9 July 2015, http://www.cetri.be/Nicaragua-is-Not-for-Sale?lang=fr.
287. Lu and Silva, 'Imagined Borders', 435.
288. Shade, 'Sustainable Development or Sacrifice Zone?', 773.
289. Ordóñez et al., *Sharing the Fruits of Progress*, 9.
290. Kennemore and Weeks, 'The Elusive Search for a Post-Neoliberal Development Model in Bolivia and Ecuador', 279; Ray and Chimienti, *A Line in the Equatorial Forests*, 34.
291. Bull and Aguilar-Stoen, *Environmental Politics in Latin America*, 2.
292. Ordóñez et al., *Sharing the Fruits of Progress*, 37.
293. Davidson and Kiff, 'Global Carbon-and-Conservation Models, Global Eco-States?', 5.
294. United Nations Agenda 2030, *Transforming our World*, Declaration, para. 18, para. 13.
295. Ibid., Declaration para. 18, para. 30.
296. Ibid., Declaration, para. 27.
297. S. Sawyer and T. Gomez, eds., *The Politics of Resource Extraction: Indigenous Peoples, Multinational Corporations, and the State*. United National Institute for Social Development (New York: Palgrave Macmillan, 2014), 6.
298. Short et al., 'Extreme Energy, "Fracking" and Human Rights', 2.
299. Arsel and Angel, '"Stating" Nature's Role in Ecuadorian Development', 208.
300. Ordóñez et al., *Sharing the Fruits of Progress*, 37.
301. S.E. Merry, 'Transnational Human Rights and Local Activism: Mapping the Middle', *American Anthropologist* 108, no. 1 (2006): 38–51.
302. M. Goodale and S.E. Merry, *The Practice of Human Rights Tracking Law Between the Global and the Local* (Cambridge: Cambridge University Press, 2007), Introduction, 35, http://humanrights.uconn.edu/wp-content/uploads/sites/767/2014/06/MarkGoodaleHumanRights.pdf.
303. Ibid.
304. Becker, 'The Stormy Relations between Rafael Correa and Social Movements in Ecuador', 50.
305. Kennemore and Weeks, 'The Elusive Search for a Post-Neoliberal Development Model in Bolivia and Ecuador', 278.
306. Becker, 'The Stormy Relations between Rafael Correa and Social Movements in Ecuador', 45.

307. The inclusion of the rights of indigenous peoples in the sustainable development agenda has been debated. See also 'Why are Indigenous People Left Out of the Sustainable Development Goals?' (Glennie, 2014); and 'Indigenous Peoples Must Not Be Left Behind upon Launch of Sustainable Development Agenda says Secretary-General at International Day Commemoration' (*UN News*, 2015).
308. United Nations Agenda 2030, *Transforming our World*, Declaration, para. 18, para. 23.
309. Ibid., para. 14.
310. World Bank, *Ecuador Overview: Context*. See also United Nations, 'UN Expert: Ecuador's Indigenous People Lack Adequate Access to Social Services', *UN News Centre*, 5 May 2006, http://www.un.org/apps/news/story.asp?NewsID=18366&Cr=Ecuador&Cr1#.V7zfGm Vlk_U.
311. UNDP, 'Ecuador: HDI Values and Rank Changes in the 2014 Human Development Report', *Explanatory Note on the 2014 Human Development Report Composite Indices* (2014), 3, http://hdr.undp.org/sites/all/themes/hdr_theme/country-notes/ECU.pdf.
312. Ordóñez et al., *Sharing the Fruits of Progress*, 7.
313. United Nations Agenda 2030, *Transforming our World*, Declaration, para. 18.

# State-led extractivism and the frustration of indigenous self-determined development: lessons from Bolivia

Radosław Powęska

This article discusses the incorporation of human rights dedicated to indigenous peoples and the problems associated with their genuine implementation in Bolivia in the context of state-led extractivism. Through this case study I will analyse the role of state and other related internal factors impacting the viability of indigenous rights related to self-determination and self-determined development. I concentrate on the problem of the character of state that can be seen as the most fundamental obstacle in implementing rights favourable to indigenous peoples' self-determined development, especially in terms of political culture, as well as historically developed state–society relations. The question of asymmetries of power and inequalities is strictly related to the 'state problem'.

## 1. Introduction

Latin America is experiencing the explosion of extractivism that results in a new imperialism for extracting natural resources, pillaging the commons and degrading the environment. This phenomenon is fuelled by the unprecedented global commodities boom.[1] The coincidence or overlapping of indigenous territories and natural resource rich areas makes indigenous peoples become a sector of society especially affected by the extractivism, gaining few or null benefits but experiencing all the harms provoked. They include devastation of lands and loss of other means of subsistence,[2] and damage of culture and social cohesion, what has even been labelled as 'cultural genocide'.[3]

It is widely accepted that in the international system there are questions that cannot be managed individually by sovereign states and these encompass such fundamental themes as environmental crisis or massive violations of human rights. For effective enforcement of the rights of indigenous peoples worldwide, the

> state cannot treat this problem alone, but its role is at the same time crucial to guarantee [international] agreements in the domestic level. The state is the entity which should put into practice (...) the recognition of indigenous peoples rights.[4]

We face a fundamental paradox – even if indigenous rights are being strengthened through international activism at the global level, their implementation strictly depends on local circumstances.

Therefore, the state tends to play a crucial role of intermediary in the dialectics of local and global levels of struggle for indigenous rights. However, apart from its role in enforcement of international human rights standards, the very same state has a voice, through its delegates, at international forums that shape human rights regulations. Furthermore, the state has a vital interest in the field of natural resource development, because in the competition over territorial and resource control (which is crucial in social-environmental and extractivist conflicts) the state remains an active player in conflicts. For Zhouri, the state plays a role of the 'meta-mediator' that

> acts as manager of the national territory, including natural resources, at the same time as it acts also as mediator of interaction between different sectors of society – firms and affected peoples – and the natural resources of the country.[5]

The ambivalent role of states in extractivism is evident. In the face of economic pressure and the increasing importance of extractivism for many national economies, the state is a custodian of legal and public order and for the security of investments, opening up regulatory frames to investors.[6]

The role of the state can thus range from being a defender and guarantor of the rights and interests of society and local communities affected by development enterprises through acting as a power and rights broker and agent of law enforcement, guarding the legal and institutional order (crucial for effective implementation of human rights). Finally, it acts as an active promotor of extractivism and ally of transnational capital to the detriment of its own citizens, with state actors turning perpetrators or becoming complicit in violating indigenous rights. In Latin America the active involvement of some states in extractivist activities is accepted because of 'the interest of the nation' or the 'majority', or in the 'public interest'.[7] The case of Bolivia acutely reflects the ambiguous and contradictory roles the state plays in the era of post-neoliberal extractivism.[8]

Sometimes rare and rather apparent victories of indigenous peoples' mobilisation against extractivist enterprises can raise exaggerated academic enthusiasm about emerging revolutions in state–indigenous peoples' relations. For example, Manuela L. Picq maintains that indigeneity and related rights to self-determination, autonomies and consultation question the state's authority over land and resources control. She argues that 'the politics of contestation against state exploitation calls for alternative sites of governance (…) In that sense, Indigenous claims to consultation challenge the authority of states over natural resources (…) Indigeneity thus defies core epistemological foundations about power'.[9]

Such statements sound very promising, but although indigenous peoples indeed defy states' authority and power, states still hold the best cards in dealing with this challenge. The state preserves all the prerogatives for shaping the legal framework regulating territorial and resources control, their use and exploitation, as well as about the effective implementation of domestic and international rights. Despite some academic optimism about indigenous movements activism on the international fora and its effect on the reconstruction of power relations within Latin American states,[10] there remains the ever-present problem of the 'implementation gap', as coined by Rodolfo Stavenhagen.[11] Any international legal provision will remain a dead letter if it is not put into operation by the state. Even in the case of an international court's judgment being unfavourable to state interests, it is the state that has the last word regarding its compliance. The case of Bolivia proves that even the ratification of well-constructed international law and incorporation of fundamental indigenous rights into the state's constitution cannot guarantee and ensure their effective realisation in practice, when confronted with complicated nuances of internal politics and development dilemmas. Rather, the opposite, Bolivia demonstrates how indigenous rights and the indigenous agenda are being deformed, manipulated and domesticated by state. No matter if the government is rhetorically pro-indigenous and itself has its roots in organic indigenous movements and maintains strategic alliances with some of them.

The election of Evo Morales as Bolivia's first indigenous president has launched the process of inner decolonisation of the state. Bolivia has incorporated one of the most advanced and far-reaching set of indigenous rights in Latin America, above all the right to self-determination, territorial autonomies, prior consultation and self-determined development. Indigenous self-determined development, understood as non-imposed forms of development in accordance with own aspirations and needs, as well as with cultural identities, traditions and institutions of indigenous peoples is not possible without autonomous decision-making. The indigenous right to freely pursue their economic, social and cultural development is inherent from their right to self-determination.

However, despite Morales' reputation as a defender of Mother Earth, by focusing on extractive sectors as a source of state revenues and supply for social redistribution through ambitious social programmes, the Bolivian state brings into question the authenticity of its pro-indigenous agenda. The extractivist priority policy quickly contradicted the official policy of *vivir bien*. The project of decolonisation became problematic for the policy of nationalisation of resources. There is a fundamental conflict between the state and many indigenous groups over this question. The expansion of hydrocarbons exploitation and mining as well as the development of infrastructure and energy projects progress at the expense of the most fundamental indigenous rights. The Bolivian state's 'pragmatic retreat' undermines indigenous rights to territorial and resource control, especially through prior consultations. It seems that the promise of the plurinational state has been converted into empty rhetoric.

This article discusses the incorporation of human rights dedicated to indigenous peoples and problems with their genuine implementation, seeking the answer to what extent those human rights can be an effective tool against extractive enterprises harmful to the interests of indigenous peoples, as well as the very relationship between extractivism and the employment of human rights in Bolivia. Through this case study I analyse the role of the character of the state and other related internal factors impacting the viability of indigenous rights related to self-determination and development. I focus especially on the political culture and historically developed state–society relations, based upon and reflecting the asymmetries of power and inequalities. These are strictly related to the 'state problem' that can be seen as the most fundamental obstacle in implementing rights favourable to indigenous peoples' self-determined development. The article begins with a brief presentation of an approach to the state theory I apply to develop my main argument – that the Bolivian state fails to protect indigenous rights despite its promises and indeed promotes extractivism instead, because of the central role of resource exploitation in generation of rents that fuel paternalist–clientelist state–society relations and help to reproduce power structures. Second, I move to theoretical discussion of indigenous self-determined development. This helps to better understand one of the central notions used in the article and to demonstrate why and to what extent the ongoing Bolivian development policy and the state's approach to indigenous peoples deeply contradict their rights. The next section describes in detail the contradictions of the Bolivian state's policy. Finally, the fourth section provides the analysis of the Bolivian 'state problem'.

I follow Abrams' assumption of state as a structure of power, based upon power relations and practices.[12] Although the state can dissimulate its non-neutral character, it is never culturally and politically neutral. Legal system norms and procedures are part of power relations and reflect the current configurations of political and social forces and their interests. However, I doubt the instrumentalist approach to state theory that suggests a state works as almost independent machine in the hands of a domestic ruling class that controls the state apparatus in order to secure its interests and maintain its domination over the rest of society.[13] The enumeration of the state's roles opening this introduction could give the impression of a very effective apparatus with unlimited powers to create and enforce its policies, albeit my intention was only to stress the importance of the state's official prerogatives that give it enormous advantage over social actors. The instrumentalism is not a very plausible approach to analyse the Bolivian case. Accordingly, we could expect that after assuming power and overtaking the whole state apparatus the Movement Towards Socialism (MAS), recognised as a genuine 'political instrument' of the indigenous-popular social movements block, would apply more counter-hegemonic economic and development policy.

The structuralist approach seems more adequate in this case, since it pays much attention to structural conditions of the state's functioning and different kinds of constraints to state's capacities, operating at international and domestic levels of interaction – for example, the pressure of neoliberal global capitalism to make peripheral countries exploit their natural resources and prioritise the profits of multinational companies. However, this approach cannot be treated as comprehensively appropriate if it suggests that a national political force ascending to power never has anything to say against capitalist domination, because the problem lies in the overall social structure of capitalist societies (fundamental relations of production) and any change of cadres controlling the state apparatus cannot change much.[14] The recent case of El Salvador can be very informative in this regard. After several years of popular protests in 2009 the pink tide government declared moratorium on mining along the Gold Belt, the richest gold area in Central America.[15] This case proves that it is possible for a poor country to oppose the interests of neoliberal multinationals and stand for the interests of society. It suggests that the impact of global capitalism pressure is not necessarily so overwhelming, unconditional and inescapable as some would see, and that the problem is more complex and nuanced.

I find Bob Jessop's strategic-relational approach (SRA) more suitable for state analysis, because it preserves all the advantages of the structuralist approach, while keeping enough space to accommodate the agency of political and social actors, whose actions are not fatally determined. Jessop argues that 'structures are strategically selective rather than absolutely constraining, there is always scope for actions to overflow or circumvent structural constraints'.[16]

The SRA stresses that state and state power are socially embedded, and 'it is precisely in the articulation between state and society (…) that many of the unresolved problems of state theory are located'.[17] Jessop sees state power as 'a complex social relation that reflects the changing balance of social forces in a determinate conjuncture'; a relation mediated through the instrumentality of juridico-political institutions, political organisations and state capacities. The state apparatus may privilege some actors, identities and strategies over others. State structures have different impacts on the ability of particular political/social forces to pursue their 'interests and strategies in specific contexts through their control over and/or (in) direct access to these state capacities'.[18]

It is clear then that asymmetries of power in a society are a crucial element here. The state system is the site of competition between the interests of diverse groups differently structurally oriented toward state power. Jessop makes it quite clear when he defines state apparatus 'as a distinct ensemble of institutions and organisations whose socially accepted function is to define and enforce collectively binding decisions on a given population in the name of their "common interest" or "general will"'.[19] But

[w]hatever the political rhetoric of the "common interest" or "general will" might suggest, these are always "illusory" insofar as attempts to define them occur on a strategically selective terrain and involves the differential articulation and aggregation of interests, opinions, and values. Indeed, the common interest or general will is always asymmetrical, marginalizing or defining some interests at the same time as it privileges other. There is never a general interest that embraces all possible particular interests (...) a key statal task is to (...) [manage] contradictions, crisis-tendencies, and conflicts to the benefit of those fully included in the "general interest" at the expense of those more or less excluded from it.[20]

Crucially, Jessop stresses that the effectiveness of state capacities depends on resources, conditions, forces and powers existing and operating beyond the state's formal boundaries; they 'depend on the structural relations between the state and its encompassing political system (...) and on the complex web of structural interdependencies and strategic networks that link the state system to its broader social environment'. It requires that the state operations 'be related both to their broader social context and to strategic choices and conduct of actors in and beyond state'.[21]

The SRA's attention to the meso-level (state–society relations) of political and social conflicts is useful for my approach that incorporates the analysis of internal factors like state–society relations' character and political culture. To be understood well, I do not want to blame the exploited world periphery instead of the exploiting capitalist centre. Neither me nor the model that I draw on to arm my arguments ignore the constraining impact of outer economic structures and world capitalist pressure. Jessop expresses it clearly enough, when adding that after Offe and Poulantzas he conceives state capacities as 'themselves dependent for their effects on links to forces and powers beyond the state'.[22] But the recognition of the role of external forces does not mean that internal mechanisms of state–society relations should be overlooked. While admitting the impact of the neoliberal world economy, I maintain that the state's internal factors mediate, and in consequence also condition, the outcome of this impact – they determine its degree, intensity and character. Thus, I propose rather a complementary and not alternative explanation. In a similar manner, I conceive the paternalist–clientelist state–society relations in Bolivia (reproducing themselves from resource revenues) as not contradictory but fully complementary with and functional to the global capitalist system, thus facilitating the subjugation of Bolivia to the interests of global resource markets.

Consequently, my approach supports the argument that because of the Bolivian state's stronger involvement in socio-political mechanisms produced/conditioned by the resource extraction-dependency and the need to respond to these mechanisms of state–society relations (state-owned resources are used to secure the political loyalty of society and the prolongation of power), Bolivia is more prone to capitulate before market pressure for natural resource exploitation. It also contributes to understanding why the Bolivian state, controlled by a supposedly pro-indigenous government, is more sensitive to the interests and expectations of some social sectors (that are structurally better situated in the system of power, thus, also more influential on the state), while at the same time it is so firm towards, relatively, the weakest part of society, who are indigenous peoples protesting against state-led extractivism.

## 2. Theorising self-determined development

The struggle for the real recognition of indigenous self-determination is at the forefront of social-environmental conflicts (conflicts over the use, control and exploitation of natural resources) with the Bolivian state, as the question of self-determined development bears also the fundamental questions of true democracy and citizenship.[23] As Salman points out, what many sectors of Bolivian society and especially indigenous communities expected while voting for Evo Morales and his MAS, was a new political culture that would go further beyond the elitist-liberal-representative democracy and seek more innovative and participatory forms of democracy recovering 'demodiversity' (coexistence of different models of democracy, RP), participation and dialogue.[24]

In this part I discuss the theory of indigenous self-determined development in order to provide appropriate context for understanding the contradictions of the Bolivian policy. I place central emphasis on the question of local subjectivity in determining own visions of development, a crucial element for indigenous communal democracy.

Going against conventional understanding of development as a unidirectional process of material progress, Amartya Sen proposes development as a process that secures possibly the greatest scope of freedom, understood as opportunity of free choice of different ways of living. For Sen, wealth is valuable as far as it allows for keeping a desirable way of living and when economic growth does not meet this condition then basic elements of social well-being are not secured. For this goal to be achieved, local communities need to have effective say on their life, which means that some kind of participative democracy is indispensable.[25] Sen also claims that 'dissent' constitutes the fundamental element of the right of people to decide about their desired development.[26] Doyle and Gilbert draw on this claim, arguing that the indigenous peoples' right to disagree with proposed development projects having impact on their territories 'should not be viewed as an obstacle to development, but as an intrinsic part of the development process and an end objective of development. The denial of this right to dissent is the denial of development itself'.[27]

The forms of development should be freely chosen and discussed because, as maintained by Escobar, in a given society, the dominating definition or vision of development is nothing but a product of power relations. As far as power relations can be changed, given society would appraise alternative kinds of development or even an 'alternative to' development. Escobar underlines the importance of social movements and their central role in constructing a counterpower that would challenge the dominant structures and visions, creating an opportunity for existence of another type of social relations based upon distinct values. Among the most important social movements are those seeking the defence of their territories as a site to enact and sustain alternative ways of living to those of the dominant paradigm.[28]

Thus, conceptions and expectations about development ought to be determined locally by and related to concrete communities living in a given place. The role of territory, identity and culture of place in alternative development options is not to overlook. The community-based social movements draw on their cultural identity and on their everyday relation to place. Place creates a relation to own territory and constitutes the main basis for collective identity, and simultaneously is the core point of reference in the complex interplay between global and local forces.[29] As Lisocka-Jaegermann concludes, the profound relations that link us to place, being an important element of local identity, contribute vastly to (and as identity is considered an essential condition for) local actions and wide mobilisation of social actors for development activities.[30] Friedmann also stresses the link between territory and democracy, turning attention back to the question of 'empowerment' of local populations and their competence of decision-making. The particular place is important, because civil society mobilises easily around local problems, and therefore it spurs a territorially organised population to seek greater autonomy and thus to achieve its own political empowerment (participatory democracy), which becomes crucial for alternative development.[31]

Locality, territoriality and a direct democracy are considered as elemental factors of 'endogenous development' or 'alternative development' as it is conceptualised by Velt-meyer. Its essential elements are: integral/holistic understanding and treatment of social, economic, cultural and environmental factors; development embedded in the realities of the local community, who should be considered as beneficiary of it; development goals are not submitted to state policy nor market mechanisms, community considers them, but shapes its development strategy according to its own aspirations; development activities are most important at the local level, therefore local communities should have control of them; engagement of local actors is an indispensable condition for development projects to be beneficial to them; subjectivity and real participation in decision-making processes, with empowerment as the basis.[32]

Direct decision-making should thus be seen as a crucial factor that enables local communities to be not only objects of external impacts, but also to develop their own development strategies, to be able to fully take advantage of their 'development repertoires', as Ray coined it. He argues that in endogenous development local people design their development vision, drawing not only on territory and resources, culture and traditions, but by functioning in constant interaction with national and international institutions and their policies. All these elements and levels form a 'development repertoire'. Extralocal regulations (including rights) can work as a support through a raising of consciousness within the local territory and by energising a cultural and political process that can allow for an emergence of 'alternative paths of development' to pursue.[33] However, looking at the other side of the coin, we can add that, for example, domestic law can be constraining and harmful in this regard. As it is shown in the next section, the Bolivian case exemplifies it very well.

In the context of Latin American, including Bolivia, natural resource development and extractivism conflicts

> arise when the meaning and the use of a certain territory by a specific group [or state actors, RP] occurs to the detriment of the meanings and uses that other social groups may employ for assuring their social and environmental reproduction,

and as a result 'the unequal access to natural resources as well as the unequal distribution of environmental risks sow the seeds of conflicts'.[34] The expansion of extractivism in indigenous peoples' territories, he argues, is a

> sort of competition between opposed geographic projects: one of successive, everyday territorial changes that are marked with historical continuities, and another project that implies drastic territorial changes, not well understood by local people and that brings greater risk and uncertainty, under the excuse of promoting modernity.[35]

Such conflicts are thus a war where the dominating conceptions are imposed upon local visions of environment and development. As crucially stressed by Bebbington, these conflicts are about a 'production of territory': about the prevailing type of relations of community with its environment and resources; who should govern territory and how; and what kind of link territory should have with its surroundings.[36]

The question of environment is without doubt an essential issue in indigenous struggles. But as Damonte Valencia explains, 'the protest of communities against environmental destruction is, in a certain way, a response to and about environmental destruction, but the social-environmental struggles have converted also into a mechanisms to gain control about fundamental aspects of communal sovereignty'.[37] As I show in the next section, in fact the indigenous peoples 'do not reject all natural resource development; rather, they seek *methods that respect their rights*, that are *consensual* and from which they can *benefit fairly*' (my emphasis, RP).[38] The main concerns include also the ways in which projects are undertaken, questions of transparency, citizenship and democracy, and problem of disregarding communities' interests.[39]

## 3.  Contradictions of plurinational extractivism

Despite the high hopes of the Bolivian indigenous peoples and international fame of Evo Morales as the first indigenous president of that country, the political practice of the state regarding indigenous rights proves to be very ambiguous and contradictory. The Bolivian government, by many called a 'government of the social movements' or an 'indigenous government' initially presented a highly positive discourse on indigenous rights. However, as I show in this section, in practice, state-led extractivism prevails over indigenous rights because the state not only fails to secure them but indeed promotes extractivism. I maintain that the crucial role of resource exploitation for the national economy (and consequently, for state power and state–society relations, see section 4), makes the Bolivian state disregard the interests of indigenous peoples affected by extractive enterprises. The imposition of fossils exploitation-based development on indigenous communities is a negation of their rights to self-determination.

The initial pro-indigenous position of Morales' government was backed by the new constitution of the Plurinational State of Bolivia (new official name of the state), promulgated in 2009. It emphasises numerous indigenous political, cultural and developmental rights, among them the right to territory, territorial-administrative autonomies, own political systems, control of renewable resources and to decide their way of development. Last but not least, the new Bolivian constitution guarantees the right to free, prior and informed consultation, obligatory for the state in any case of legislative acts or administrative decisions that would affect the interests of indigenous peoples, including the exploitation of non-renewable resources from their territories (Art. 30, 352, 403).[40] Such recognition follows the path established by the ratification of Convention 169 of the International Labour Organisation (ILO 169) in 1991 and the UN Declaration on the Rights of Indigenous Peoples (UNDRIP). These international human rights instruments are of enormous importance for indigenous peoples' claims to self-determined development. Bolivia not only ratified UNDRIP in 2007 but the same year converted it in its entirety into the national law (National Act nr 3760),[41] being the only country in the world to do so. Therefore, any discussions of whether the UNDRIP as an international declaration is binding or not are pointless.

ILO 169 establishes that indigenous peoples

> shall have the right to decide their own priorities for the process of development as it affects their lives, beliefs, institutions and spiritual well-being and the lands they occupy or otherwise use, and to exercise control, to the extent possible, over their own economic, social and cultural development. In addition, they shall participate in the formulation, implementation and evaluation of plans and programmes for national and regional development which may affect them directly (Art. 7).

The convention also requires governments to 'consult the peoples concerned, through appropriate procedures and in particular through their representative institutions, whenever consideration is being given to legislative or administrative measures which may affect them directly'. And what is especially crucial, 'the consultations shall be undertaken 'in good faith and in a form appropriate to the circumstances, with the objective of achieving agreement or consent to the proposed measures' (Art. 6). Regarding the situation when the subsoil resources pertaining to indigenous peoples' lands belong to states, the 'peoples concerned shall wherever possible participate in the benefits of' exploitation activities, and 'shall receive fair compensation for any damages which they may sustain as a result of such activities' (Art. 15). Moreover, 'In applying national laws and regulations to the peoples concerned, due regard shall be had to their customs or customary laws' (Art. 8).[42] This is important regarding the problematic question of collective land titling and recognition of indigenous territories, issues that raise never-ending conflicts between national Bolivian regulations and indigenous peoples' vision and traditions of their understanding of territories, as will be shown in further paragraphs.

The UNDRIP goes further. First of all, in its Article 3 the document states that 'Indigenous peoples have the right to *self-determination*. By virtue of that right they freely determine their political status and freely pursue their economic, social and cultural development.' The next article affirms that 'Indigenous peoples, in exercising their right to self-determination, have the right to autonomy or self-government in matters relating to their internal and local affairs' (Art. 4), and they have the right to 'maintain and strengthen their distinct political, legal, economic, social and cultural institutions' (Art. 5). Indigenous peoples have 'the right to participate in decision-making in matters which would affect their rights, through representatives chosen by themselves in accordance with their own procedures, as well as to maintain and develop their own indigenous decision-making institutions' (Art. 18). Art. 20 reaffirms indigenous peoples' right to 'maintain and develop their political, economic and social systems or institutions, to be secure in the enjoyment of their own means of subsistence and development, and to engage freely in all their traditional and other economic activities'. Finally, Art. 23 stipulates that 'Indigenous peoples have the right to determine and develop priorities and strategies for exercising their right to development'. They have the right 'to be actively involved in developing and determining (…) economic and social programmes affecting them and, as far as possible, to administer such programmes through their own institutions'.[43]

The UNDRIP as a whole explicitly supports indigenous claims to self-determination and self-governance in the very clear context of development, strengthening indigenous rights to self-determined development. This is further extended by the rights to consultations. UNDRIP Art. 19 requires states 'to consult and cooperate in good faith with the indigenous peoples concerned through their own representative institutions in order to obtain their free, prior and informed consent before adopting and implementing legislative or administrative measures that may affect them'. Then the document confirms this right in the context of development projects and extractivism. 'Indigenous peoples have the right to determine and develop priorities and strategies for the development or use of their lands or territories and other resources'. Moreover, states shall seek indigenous peoples' 'free and informed consent prior to the approval of any project affecting their lands or territories and other resources, particularly in connection with the development, utilisation or exploitation of mineral, water or other resources' (Art. 32).[44]

The issue of whether the requirement of obtaining free, prior and informed consent (FPIC) embodies the right to dissent or a so-called veto right (to withhold agreement), is highly debated (especially, states and companies refuse to accept stronger interpretation), but many International Human Rights (IHR) analysts support the indigenous right to say no to proposed development projects that may affect them.[45] For Imai '[s]elf-determination refers to a choice, not a particular institutional relationship'.[46] In a similar manner, the International Court of Justice defined the principle of self-determination as 'the need to pay regard to the freely expressed will of peoples'.[47] The right of indigenous peoples to refuse proposed activities can be seen as a natural (logical) extension of the right to consultation,[48] because there cannot be consent if there is no possibility of disagreement. As Doyle and Gilbert point out, the lack of possibility for non-consent 'implies that communities have no determining say in the developments that occur on their own lands and (...) cannot determine their own development priorities',[49] what contradicts numerous rights related to indigenous self-determined development and self-determination codified in the IHR documents.

Ten years since Morales' first electoral victory and despite the early ratification of the important IHR instrument ILO 169, and over seven years since the promulgation of a progressive new constitution, Bolivia still lacks real recognition of self-determined development rights and especially the right to consultation under the principles of FPIC. The constitution mentions previous consultations several times (Arts 30, 352, 403) but never admits that the goal should be consent. However, in Art. 11 the constitution mentions consultations with indigenous peoples as a form of direct and participative democracy that constitutes the Bolivian political system.[50] The problem with the implementation of indigenous rights is directly related to the new economic policy launched by the government in 2006. It is based on the central role of the state in the economy, modernisation projects and nationalisation of strategic resources (mainly gas alongside oil and the more traditional mining industry), being treated as an 'engine' for the country's development. The export revenues provide a budget for public investments and industrialisation projects, but they also fuel the system of redistribution through state-managed social programmes, such as the *Renta Dignidad* (Dignity Pension) programme for seniors, Juancito Pinto stipend for children who regularly attend school, and the Juana Azurduy payments to pregnant women and mothers of infants.[51] We need to acknowledge the great achievements of new state policy in terms of economic growth (GDP per capita in 2005 was US$1,018 in 2005 and US$3,147 in 2014) and poverty reduction. In 2005, 60% of all Bolivians were poor and 36.7% were extremely poor. In contrast, in 2013 the level of poverty fell to 39% and extreme poverty level dropped to 18.9% (in rural areas the total poverty level was 80% in 2005 and 60% in 2013, while extreme poverty dropped from 65.6% in 2005 to 38.8% in 2013).[52]

The centrality of resource extraction to the national economy cannot be underestimated. In the last few years the gas and oil industry, together with traditional mining, generated 70% of all export value in 2013.[53] The overall area designated to gas and oil exploitation has grown eightfold since 2007 and in 2012 it was about 24 million hectares, about 22.5% of all the national territory. This included 64 indigenous territories affected by subsoil resources extractivism (37 in the lowlands and 27 in the highlands).[54] In 2010 the president issued Supreme Decree 0676 conceding vast territories of protected natural areas for hydrocarbon exploitation.[55] This move was repeated in 2015 with Supreme Decree 2366 that opened up other extensive protected areas for extractivism.[56] Moreover, Bolivia became interested in fracking its shale gas reserves, which are estimated to be 48 trillion cubic feet (tcf), the fourth greatest in Latin America.[57] Yet in 2013 the Bolivian government authorised a technical study about the prospects for fracking. Up to now the programme of shale gas exploitation has not materialised, but given the growing appetite of the state for resource export revenues, it is possible that it could be enforced in the near future.[58]

The importance of resource extractivism for the national economy exacerbates the tendency to accelerate the consultations or omit them, in order to give way fast to exploitation with the lowest costs and the biggest volume possible, avoiding prolonged deliberations and negotiations. For the state the consultations are a waste of time and only complicate effective economic and social policy. But this problem does not refer only to subsoil exploitation. The most infamous case of indigenous rights violation and subsequent conflict of international fame had to do with the government's imposition of a decision to build a highway through the Territorio Indígena Parque Nacional Isiboro Sécure (TIPNIS), a national park and indigenous territory. In 2011 the inhabitants organized a march to the country's capital, demanding dignity, respect for territories, compliance with consultation right, suspension of construction works and all the hydrocarbon exploitation activities in the national parks, and realisation of indigenous autonomy and participation in all development projects (resource exploitation, hydroelectric dams and road construction) that may affect their communities. The protest was brutally broken down by the police.[59]

Bolivia does not have any general law regulating indigenous consultations. The existing proposal for a Prior Consultation Law was ready in 2014, but it never entered into parliamentary debate and is awaiting a 'better future'.[60] The proposed law very carefully and evasively establishes that the aim of consultations is to obtain consent, but its general overtone reveals a very superficial and manipulative understanding of consent, in no way anticipating the actual withholding of consent by the affected population. The proposed law mentions consent only as a 'consensus achieved in the process of intercultural dialogue' between the state and indigenous peoples, and limits its scope to the incorporation of the population's view of development into the implementation of administrative decisions. Moreover, the project literally 'guarantees the execution and continuation of extractivist activities' given their 'strategic character and public interest for the national development'.[61] The lack of any enforced regulatory law about consultations in activities other than resource extractive ones, for example, construction of dams and other infrastructural projects, is alarming and will likely provoke new conflicts in the future. It is widely acknowledged that dams, as they are used to extract water and hydro-power, nearly always bring about social and environmental costs.

Up to now only a few laws and decrees regulating activities in particular sectors of the economy have been 'armed' with the specific codification of the consultation right. It is surprising that with mining being so important for the Bolivian economy throughout history and having had such a huge impact for all the highland parts of Bolivia, the regulations for consultation in mining were implemented only in 2014. The lack of regulations provoked several claims for damages on the part of indigenous communities, one of the most famous being the Coro Coro case around the mine which devastated pastoral lands and sources of water for its local community and its operation had never been the subject of a consultation nor an environmental study. The case was brought to the Inter American Human Rights Commission and the UN Permanent Forum on the Indigenous Issues. The new Mining Law explicitly denies indigenous rights to dissent to mineral activities and allows only for negotiations about mitigation of impacts and compensation.[62]

In many cases indigenous peoples affected by extractive activities complain about the lack of any consultations. One of the biggest problems regarding the right to consultation in Bolivia is related to the issue of collective land titling and legal recognition of indigenous territories. The state does not apply the consultation mechanism in affected areas where the indigenous population lack legal territorial recognition. This is especially worrying among the Guaranís, the most impacted by hydrocarbon exploitation. Despite the fact that legal frameworks regulating indigenous territorial recognition have existed in Bolivia since 1996, the entitlement processes are extremely slow and there are many cases when the processes cannot be concluded for decades, depriving local populations of their rights.[63] The latest case occurred in Takovo Mora territory, invaded by a gas exploitation project without consultation of the local Guaraní people. Tired of the endless administrative struggle (the Guaranís confirm that they were asking for entitlement for 19 years), in 2015 the indigenous people organised a road blockade to demand compliance with their rights. The protest was brutally repressed by the police.[64]

The condition of legal entitlement of territory for the execution of the consultation right is considered a severe violation of human rights in light of existing interpretations of indigenous peoples' rights by the Inter American human rights system. In 2001, in the case of the *Mayagna (Sumo) Awas Tingni Community* v. *Nicaragua*, the Inter American Human Rights Court (IAHRC) declared 'the indigenous people's property rights originated in indigenous tradition and, therefore, the State had no right to grant concessions to third parties in their land'. The court decided that 'the State had to adopt the necessary measures to create an effective mechanism for demarcation and titling of the indigenous communities' territory, in accordance with their customary law, values, customs and mores', and 'until such mechanism was created, the State had to refrain from any acts that might affect the existence'.[65] Later, in 2003, dealing with the case *Maya Toledo District* v. *Belize*, the court confirmed that 'the property rights of indigenous peoples are not defined exclusively by entitlements within a state's formal legal regime, but also include that indigenous communal property that arises from and is grounded in indigenous custom and tradition'.[66]

In cases when consultations are undertaken in hydrocarbon exploitation activities, several problems are being reported. One of the biggest concerns is about the lack of impartiality of the governmental entity responsible for organising, conducting and monitoring consultation processes, which is the Ministry of Hydrocarbons and Energy, obviously vitally interested in fast and problem-free authorisation of new projects. Moreover, there are numerous cases of attempts to co-opt and corrupt indigenous organisations, that can lead to internal conflicts and divisions.[67]

Another notorious problem is a conflict between norms and visions of indigenous peoples and norms and regulations used by the state, which has serious implications for the meaning of 'serious impact' on the environment and/or community and its means of social and economic reproduction. It refers also to the question of territorial and political integrity of indigenous peoples. The state consults only directly affected communities. The indigenous peoples claim that the whole greater political entity (to which affected communities belong) of particular people should be taken into account, according to the principle of self-determination and the respect for the internal political structures of a given people.[68]

Last but not least, indigenous organisations complain that the consultations involve only the issue of negotiation of mitigation of impact and the preparation of environmental impact studies. Moreover, the people receive ready-made projects to discuss and so there is no scope for collaboration with companies nor the state about project design, thus reducing the consultations to mere rubber-stamping approval. Instead of the state's good faith, there is disinformation, hiding negative side-effects of projects, manipulation of data, and propaganda about achievements.[69]

Thanks to its organic ties to social movements and the indigenous symbolism associated with the president (indigenous/peasant origin, superficial ethnic rhetoric, etc.), the government represents itself as a legitimate incarnation of indigenous power and a natural advocate of indigenous peoples' interests. This discourages dissenting indigenous groups from defending their rights and opposing the state's extractivism. The protesters are depicted as going against the common interests of all the popular sectors of society. The government and its allies say that such minorities should not be an obstacle to the development of the country and that it cannot be that some small indigenous groups are able to prevent the rest of society taking advantage of national resources. The dissenting protest groups are also portrayed as trouble-makers opposing the 'process of change', and traitors, paid off by foreign non-governmental organisations and USA.[70]

Maybe the solution to all these problems with indigenous self-determined development would be the recognition of indigenous peoples' right to veto or even recognition of the collective property of all the natural resources, including subsoil, of their traditionally occupied territories? The controversies around consultations of extractive activities would be radically reduced with the veto right and virtually disappear in the case of indigenous subsoil control. Certainly this would mean a more genuine self-determination of the indigenous peoples and a real opportunity for their self-determined development. The indigenous peoples, through their own political institutions, would deliberate internally about the best way to development, considering questions of subsoil extraction and their possible advantages and risks, together with the issue of environmental preservation. In the case of an explicit veto right, indigenous peoples would have meaningful decision-making competence, but the subsoil control would give them greater development opportunities. Not only empowerment to say no, but also to benefit directly from extractive activities. We can imagine that the revenues could be shared 50:50 between communities and as taxes to the state's budget so as not to harm the national interests. The autonomous indigenous political entities would be economically stronger and more independent (actually a somewhat similar solution already exists in the United States and it can be very cautious inspiration[71]). However, this would be perceived by local elites and global capitalism as allowing indigeneity to trump development and hence make Bolivia unattractive to foreign direct investment (FDI).

Yet during the drafting of the new constitution in 2006–2007, many indigenous organisations demanded the right of indigenous communities to own and control all of the natural resources located in their territories, including non-renewable ones and subsoil. Such claims were presented to local populations during constitutional workshops and were announced in the congress of the biggest indigenous-peasant organisation Central Sindical de Trabajadores Campesinos de Bolivia (CSUTCB), although finally this movement, a strategic ally of the government, did not include those claims in its constitutional proposal.[72] The important organisation of traditional Andean communities, Consejo Nacional de Ayllus y Markas del Qullasuyu (CONAMAQ), demanded in its constitutional project that indigenous autonomies have right to 'property and sustainable administration of renewable and non-renewable natural resources'.[73] The Asamblea del Pueblo Guarani (APG) also claimed the property and its right to administer the indigenous peoples' soil and subsoil. The Central de Pueblos Etnicos Moxenos del Beni (CPEMB) claimed that 'the indigenous peoples are owners of the renewable and non-renewable resources in their territories and have a right to administer, use, and sustainably control the resources.[74] Finally, the coalition of the nine biggest indigenous peoples organisations in Bolivia prepared a compromised constitutional project reflecting different interests of the movements. In that document the movements demanded indigenous peoples' rights to 'participate in the co-administration of the non-renewable natural resources with the national government', to 'participate in direct, real and effective form in administration, management, decision-making and benefits coming from the use of natural resources (…) found in their territories' and claimed that 'indigenous peoples can be partners in forming companies undertaking resource exploitation'.[75] Given the economic interests of the state and different interests of other parts of society, these were utopian claims.

### 4. Compensatory or predatory? Problem of the state and asymmetries of power

At the very beginning of the formation of the new government and new state project in Bolivia, vice-president Álvaro García Linera defended his concept of 'Andean-Amazonian capitalism' to be constructed in the frame of 'decolonisation of the State', as based upon

[t]he construction of a strong State that would regulate the expansion of the industrial economy, extract its surplus and transfer it to the communitarian sphere in order to strengthen and foster forms of self-organisation and truly Andean and Amazonian economic development, avoiding that 'the modern' steal and use up all the energies from 'the communitarian', promoting its autonomous development.[76]

In fact, as it is shown further, the state's ongoing policy contradicts what had been declared a decade ago. In this section I analyse the state's features contributing to this situation.

In Bebbington's view of the Andean (neo)extractivism,

what is being set up is a cultural logic and a form of occupation and control of space that reflects the power of the centre over regional and local spheres, the non-indigenous and urban power over indigenous-peasant groups, and the power of [external, private, state or mixed, RP] investments over [local, RP] collective institutions. It is in some way a culmination of the *internal colonisation* (my emphasis, RP).[77]

It is true that Bolivia is a state with a long colonial legacy, with a tradition of discrimination and exclusion of indigenous peoples. Historically it was a country where privileged sectors used state power as a mechanism to secure exploitation of subaltern groups.

I would also add that what we face in Bolivia is in my opinion a perverse variant of Harvey's 'accumulation by dispossession'.[78] While this concept is widely employed to critical studies of neoliberalism (it describes centralisation of wealth and power in the hands of a few by dispossessing the public of their wealth or land), I propose to invert its meaning. I use it to address the indigenous peoples' resource dispossession by central state power as transferring property from indigenous groups within a state to state ownership for the benefit of domestic elites and other social groups that are favourably oriented towards central power. Such transfer is made under the figure of redistribution of resource export revenues, but also through state investments that directly benefit entrepreneurial sectors. Thus Bolivia resembles a predatory state. This kind of state is described as one which in the interests of the best-organised groups deprives society of resources or plunders surplus without reward for the welfare of the harmed population.[79] All of these above interpretations refer to the problem of unequal socio-political relations that render the power structure of the state.

Bolivia also demonstrates some attributes of a classic rentier state. The country is historically a natural resource extraction-dependent state. Such a state cannot function without revenues coming from resource export sectors, and it has a strong tendency to centralism and vertical relations with society. These patterns are conditioned by the need for control of vital resources and strategic sectors of the economy. Inevitably such a state would maintain a strong centralist character. This goes hand in hand with the predatory character of such a state as was mentioned above. Acosta maintains that Latin American rentier states, enjoying abundance of non-renewable resources and that are primary-export-led and rent-seeking, generate several problems with dis-incentives for long-term and diversified stable development, note extremely high levels of inequality and experience (and reproduce by themselves) serious deficits of democracy through paternalism and clientelism.[80] Acosta's arguments resemble the well-known, though disputed, findings of Ross and his influential 'Does Oil Hinder Democracy?' with arguments about the authoritarian and anti-democratic character of rentier states.[81]

The thesis about the anti-democratic character of rentier states might appear incompatible with contemporary apparently democratic Latin American states, especially in the case of Bolivia. There is a widely shared assumption that the enfranchisement and empowerment of the people, especially historically excluded indigenous sectors, are greater now than ever. But I want to stress that the question here is not about formal democracy or minimal standards of liberal representative democracy. The true concern is about the tendency for state-sponsored centralism and paternalist-clientelist relations that hinder evolution towards genuine decentralisation and participative/direct democracy, as well as the full realisation of indigenous peoples' rights.

For Bolivian researcher and analyst Marco Gandarillas, 'an evident erosion of democracy and serious violations of the most basic rights are produced proportionally to the deepening of extractivism'.[82] The limitations and obstacles to indigenous consultation described in the previous section show clearly the link between the Bolivian state's rent-seeking and the constraining of indigenous peoples' rights to decide their own development. Tockman and Cameron provide findings that support the above arguments in the context of the lack of implementation of indigenous autonomy in Bolivia. For the authors, the main barriers to autonomy are the state's effort to control natural resources and to intensify extractivism. It is clear that the state does not allow greater indigenous territorial control that would strengthen local resistance to extractive enterprises and complicate exploitation proceedings. The second reason is the quest to maintain the system of rural networks of popular support and control of political power. Indigenous autonomy would undermine partisan politics and detach indigenous territories from the MAS political patronage structure. However, Tockman and Cameron see subsoil resource control and the question of control of political power separately.[83] In my view there is an important relationship between these issues in terms of the functioning of the state's paternalism and patron–client model that regulates state–society relations. I discuss this link more below.

There could be a conceptual problem with combining this analysis (and especially concepts of internal colonialism, predatory state, etc.) with Eduardo Gudynas' 'compensatory state'. However, what Gudynas describes is a new version of the classic extractivist state with the difference being in the emphasis placed on redistributive economic justice.[84] But we need to be aware, first, that redistribution is centrally managed and controlled by the state, reinforcing its centralist character.[85] Second, the redistribution programmes are an important part of a new state's mechanisms for legitimating itself among vast parts of society, and the predatory acts of state are oriented towards weaker, more vulnerable and culturally and legally subjugated sectors – most notably, indigenous peoples. It is a mechanism of exploitation of indigenous communities by another social sector – mainly, new indigenous and *mestizo* urban middle-class and local business sectors, working hand in hand with multi-nationals.

Actually, Gudynas admits this when discussing some important aspects that complicate the relations between the compensatory state and its society. The compensatory Bolivian state cannot be neutral in benefitting society as a whole, because its revenues rely entirely on extraction and export of natural resources that are situated in specific territories occupied by specific peoples with claims over territorial and resource control in these areas. The neoliberal extractivist predatory state inevitably privileges some sectors with redistributed benefits at the expense of the population targeted by the extractivist projects. These communities are the recipient of the negative side effects (environmental damage, loss of land and resources, social divisions within the community, loss of alternative development opportunities and means of economic and social reproduction, etc.).[86]

So, is the Bolivian state compensatory or predatory? We can say it is both at once. In order to be compensatory towards dominating parts of society (or merely enriching the subaltern elite), it is simultaneously predatory towards indigenous peoples occupying resource-rich areas. But how can we explain this ambiguous nature of the Bolivian state's performance, based upon the contradiction of pro-indigenous discourse and pro-extractivist economic policy? Let us target the interplay of structural and conjunctural factors when studying the 'double face' of MAS incarnated in the current state project. The ruling party almost since its beginning combined two ideological and pragmatic (in terms of demands, expectations, and particular interests) wings or discursive axes: nationalist, anti-neoliberal, trade unionist, interested in the return of the economically active central state, the revocation of privatisation, nationalisation of hydrocarbons and redistribution of rents, industrialisation and general modernisation, and the generation of employment. The other 'wing' dealt with ethnic issues – claims for the end of the persistent exclusion and marginalisation of native sectors of society that sought greater access to and presence in the political system, greater sensibility of the state to the interests of indigenous peoples and conferring collective rights, for example, territorial autonomy, communitarian justice and democracy, recognition of cultural rights, and so on. There was no one agenda, but several different agendas that formed an unfocussed scope of interests and expectations for the state's renovation; agendas of different sectors of society that felt similarly harmed by imperialism and neoliberalism. While these different dimensions combined well before the winning of political power, the apparent union of the indigenous-popular block started to dissolve thereafter. Priority was given to the primary-export logic and economic centralism that would generate extensive rents to redistribute, and the ruling party opted for a national-popular and state-centric political project for the Bolivian 'refoundation', actually a typically developmentalist project. The legitimation of the refoundation project was based upon the rhetoric of the plurinational state, albeit only operationalised in minimalist terms.[87]

The new state's priorities reflect well the actual configuration of power and political influence of particular sectors of society. The circle of the closest allies and the most influential sectors includes not only urban indigenous and *mestizo* bourgeoisie, but also the majority of indigenous-peasant movements. It seriously hinders if not disables prospects for united struggle for genuine implementation of indigenous peoples' rights in Bolivia. This can seem surprising giving the strong ethnic character of the social upheaval that raised Evo Morales to power a decade ago. However, it should be stressed that the indigenous/peasant sectors are highly diverse. There are more traditional groups with stronger ties to their ancestral territories and carrying out more integral approaches to territory and resources. But they coexist with less traditional, less territory-tied groups, that are more market-oriented and have a different approach to resources, environmental protection and the general idea of development (especially *cocaleros* and colonisers). These sectors of indigenous peasantry are without any doubt in strategic alliance with MAS (actually they initiated the formation of MAS) and their perspectives are better incorporated in the state's policy. The obvious discrepancies in interests that exist among the Bolivian indigenous peoples fuel divergent approaches to the very theme of indigenous rights and state policy.[88]

The configuration of power and dominating social interests and expectations cannot be underestimated. Given this, the evolution of the state project towards centralism and the substantial reduction of 'plurinational' elements of the state's ideology is following traditional and well-established patterns of the Bolivian political culture and character of state–society relations. These include the well-known phenomena of vertical state–society relations and clientelist statism, strong paternalism and corporatist clientelism, all related to the problem of a rentier state. Accordingly, to Laserna, corporativism, clientelism, statism and rentierism are interrelated and constitute together the backbone of the Bolivian 'national ideology' and political system, historically rooted but re-articulated with the world commodities boom. Laserna points out that this model incites conflicts between different social groups competing to gain influence in state power and to capture rents.[89] State-owned resources are used to secure the political loyalty of different social groups and the prolongation of power. The maintenance of the system serves the reproduction of power structures and the position of ruling actors and privileged groups. Clientelism and paternalism reproducing themselves from resource revenues are not contradictory but fully complementary with global capitalism, as were the interests of the Bolivian elites throughout history.

There is a fundamental contradiction between this model of state control of resources, indispensable for the generation of rents that fuel paternalist-clientelist state–society relations, and indigenous peoples' self-determination in development. In its perverse logic of power, the state's paternalism discourages society's own initiatives. Instead of increasing incentives for people's own choices and direct opportunities of development, the state limits people's autonomy in disguise as the protector and saviour of society. Even if all listed above can be recognised as important factors contributing to the problem of corruption that regularly raises social discontent in Bolivia, as well as to the erosion of genuine, direct democracy, such a vision of state and society relations are widely accepted in Bolivian society. It is the Latin American 'country with the highest percentage of population receiving social assistance and a country with the highest acceptance of rent-redistributive policy'.[90]

Recently, Bolivia, similar to Ecuador and Venezuela, suffered a severe decline of state revenues from fossil fuels, caused by the fall in global prices. However, the government's response, instead of seeking alternatives to resource exportation, has been to further intensify resource exploitation and create a new set of incentives for foreign investment that would help to recover prosperity.[91] Such a reaction would be a proof of dependency of peripheral economies from the pressure of global capitalism, but in my opinion it can be equally a response to the necessities of a domestic model of state power and state–society relations.

## 5.  Conclusions and final remarks

In this article I have discussed the incorporation of indigenous peoples' rights in Bolivia and several serious problems with their genuine implementation. I have explored the relationship between the human rights and extractivism led and promoted by the state and have examined to what extent the human rights based approach can be an effective tool against enterprises harmful to the interests of indigenous peoples. The study of a role of the state and other related internal factors was crucial to develop my arguments. I have concentrated on the problem of the character of state that can be seen as the most fundamental obstacle in implementing rights favourable to indigenous peoples' self-determined development, especially in terms of political culture, as well as historically developed state–society relations.

The starting point is the observation of the fundamental paradox of the rhetoric of human rights and its uses and abuses by the Janus-faced state – one face compensatory and the other predatory. Even if indigenous rights are being strengthened through international activism at the global level, their implementation strictly depends on local circumstances. The state plays a crucial role of the intermediary sphere in the dialectic between local and global levels of struggle for indigenous rights. The Bolivian case provides the proof that even the ratification of well-constructed international law and incorporation of fundamental indigenous rights into the constitution cannot ensure their effective realisation in practice. Especially when confronted with complicated nuances of internal politics and development dilemmas. Bolivia demonstrates the opposite. The indigenous rights and indigenous agenda are being deformed and manipulated by the state. And it does not really matter if the government is discursively pro-indigenous. The expansion of hydrocarbons and the mining industry, together with infrastructural and energetic projects are at the expense of the most fundamental indigenous rights. This 'pragmatic retreat' undermines rights to territorial and resource control, especially through prior consultations.

The central problem in my explanation of these contradictions lies in the field of asymmetries of power and inequalities related to the 'state problem'. It is characterised by vital phenomena of a paternalist-clientelist model of state–society relations and rentierism. The undisturbed generation of rents indispensable to reproduce this model requires the imposition of the state's interests upon the most vulnerable part of society, which is indigenous peoples, and the limitation of their right to self-determination in development. The state remains the key to the real self-determination and self-determined development of the indigenous peoples but also their subordination. This forms an interesting paradox, because state power can only loosen with changes leading towards the achievement of those rights, while simultaneously the state remains the most important and powerful force that can make it possible and without which even the best regulations stay only in paper.

We can consider if a direct transfer of administrative control over subsoil and its benefits to the indigenous peoples would help to reduce existing inequalities and power asymmetries. I believe it would give an impulse to great mobilisation for empowerment and direct decision-making of the communities. Even with indigenous ownership of non-renewable resources, they would more consciously and responsibly consider different paths of development in order to choose the most promising and beneficial solutions in the long term, having in mind the question of the environment and its preservation. There would be other sorts of problems to solve, including the issue of indigenous internal inequalities, hierarchies and power asymmetries. It should be further addressed in the future. Moreover, it is possible that the explicit recognition (at the international level) of a mere veto right to proposed development projects would continue to encourage states to ignore and disregard such rights or provoke more open forms of human rights violation, what would mean a deterioration of human rights' practical standards. And the recognition of the Bolivian indigenous peoples' rights to subsoil ownership seems completely unreal because of the economic and socio-political, structural factors described in this article.

Notwithstanding these pessimistic considerations of possible changes to this situation, I believe that it is important to pay more attention to the domestic factors influencing human rights implementation, and to discuss more sincerely the fundamental problems behind these factors – inequality and power asymmetries.

## Disclosure statement

No potential conflict of interest was reported by the author.

## Notes

1. Michela Coletta and Malayna Raftopoulos, 'Whose Natures? Whose Knowledges? An Introduction to Epistemic Politics and Eco-ontologies in Latin America', in *Provincialising Nature: Multidisciplinary Approaches to the Politics of the Environment in Latin America*, ed. Michela Coletta and Malayna Raftopoulos, (London: Institute of Latin American Studies, School of Advanced Studies, University of London, 2016), 1–17.

2. Corinne Lennox, 'Natural Resource Development and the Rights of Minorities and Indigenous Peoples', in *State of the World's Minorities and Indigenous Peoples*, ed. Beth Walker (London: Minority Rights Group International, 2012), 11, 16.

3. Julian Burger, *Report from the Frontier: The State of the World's Indigenous Peoples* (London: Zed Books, 1987), 105.

4. Nieves Zúñiga García-Falcés, 'Conflictos por recursos naturales y pueblos indígenas', *Pensamiento Propio* 22 (2005): 33–61, 52 (author's own translation).

5. Andréa Zhouri, *Mapping Environmental Inequalities in Brazil. Mining, Environmental Conflicts and Impasses of Mediation* (Berlin: desigALdades.net, 2014), 9.

6. Maristela Svampa, and Mirta Antonelli, eds, *Minería transnacional, narrativas del desarrollo y resistencias sociales* (Buenos Aires: Biblios, 2009), 31.

7. Lennox, 'Natural Resource Development', 10–21, 11, 16–20.

8. On the ambivalence of post-neoliberal development and new extractivism in Bolivia and elsewhere in Latin America, see Henry Veltmeyer and James Petras, eds, *The New Extractivism. A Post-Neoliberal Development Model or Imperialism of the Twenty-First Century?* (London: Zed Books, 2014).

9. Manuela Lavinas Picq, 'Self-Determination as Anti-Extractivism: How Indigenous Resistance Challenges World Politics', in *Restoring Indigenous Self-Determination. Theoretical and Practical Approaches*, ed. Mark Woons (Bristol: E-International Relations, 2014), 26–33, 30–1.

10. See especially Alison Brysk, *From Tribal Village to Global Village. Indian Rights and International Relations in Latin America* (Stanford, CA: Stanford University Press, 2000).

11. Rodolfo Stavenhagen, Report of the Special Rapporteur on the Situation of Human Rights and Fundamental Freedoms of Indigenous People E/CN.4/2006/78 (UN, 2006).

12. Philip Abrams, 'Notes on the Difficulty of Studying the State', *Journal of Historical Sociology* 1, no. 1 (1988): 58–89.

13. Clyde W. Barrow, *Critical Theories of the State. Marxist, Neo-Marxist, Post-Marxist* (Madison: The University of Wisconsin Press, 1993), 13–50.

14. Ibid., 51–76.

15. Robin Broad and John Cavanagh, 'Poorer Countries and the Environment: Friends or Foes?', *World Development* 72 (2015): 419–31.

16. Bob Jessop, 'The State and Power', in *The SAGE Handbook of Power*, ed. Stewart R. Clegg and Mark Haugaard (London: SAGE Publications, 2009), 367–82, 375.

17. Ibid., 370, 371.

18. Ibid., 376–8.

19. Bob Jessop, *State Theory: Putting the State in its Place* (Cambridge: Polity, 1990), 341.

20. Jessop, 'The State and Power', 373.

21. Ibid., 376–7.

22. Ibid., 378.

23. Fernando Mayorga, 'Bolivia: segundo gobierno de Evo Morales y dilemas del proyecto estatal del MAS', in *La actualidad política de los países andinos centrales en el gobierno de izquierda*, ed. Yasuke Murakami (Lima: IEP, 2014), 13–54, 28.

24. Ton Salman, 'El estado, los movimientos sociales y el ciudadano de a pie: exploraciones en Bolivia entre 2006 y 2011', *América Latina Hoy* 65 (2013): 141–60, 150–1.

25. Amartya Sen, *Development as Freedom* (New York: Oxford University Press, 1999).

26. Ibid., 36–7, note 18.

27. Cathal Doyle and Jérémie Gilbert, 'Indigenous Peoples and Globalization: From "Development Aggression" to "Self-Determined Development"', *European Yearbook of Minority Issues* 7 (2008/9): 219–62, 251–2.

28. Arturo Escobar, *Encountering Development. The Making and Unmaking of the Third World* (Princeton, NJ: Princeton University Press, 1995).
29. Arturo Escobar, 'Culture Sits in Places: Reflections on Globalism and Subaltern Strategies of Localization', *Political Geography* 20 (2001): 139–74.
30. Bogumiła Lisocka-Jaegermann, Kultura w rozwoju lokalnym. Dziedzictwo kulturowe w strategiach społeczno-gospodarczych latynoamerykańskich społeczności wiejskich [Culture in Local Development. The Cultural Heritage in Social-Economic Strategies of Latin American Rural Communities] (Warsaw: Wydział Geografii i Studiów Regionalnych, 2011), 56–7.
31. John Friedmann, *Empowerment: The Politics of Alternative Development* (Oxford: Blackwell, 1992), vii.
32. Henry Veltmeyer, *The Search for Alternative Development* (Working Papers in International Development Halifax, NS: Saint Mary's University, 1996), 22–5.
33. Christopher Ray, 'Towards a Meta-Framework of Endogenous Development: Repertoires, Paths, Democracy and Rights', *Sociologia Ruralis* 39, no. 4 (1999): 522–37, 525.
34. Zhouri, *Mapping Environmental Inequalities*, 7–8.
35. Anthony Bebbington, 'Elementos para una ecología política de los movimientos sociales y el desarrollo territorial en zonas mineras', in *Minería, movimientos sociales y respuestas campesinas: una ecología política de transformaciones territoriales*, ed. Anthony Bebbington (Lima: IEP, 2011), 53–76, 54–5.
36. Ibid., 63.
37. Gerardo Damonte Valencia, 'Mineria y politica: la recreacion de luchas campesinas en dos comunidades andinas', in *Minería, movimientos sociales y respuestas campesinas: una ecología política de transformaciones territoriales*, ed. Anthony Bebbington (Lima: IEP, 2011), 147–92, 190.
38. Lennox, 'Natural Resource Development', 11.
39. Anthony Bebbington, '¿Una nueva extracción, una nueva ecología política?' in *Minería, movimientos sociales y respuestas campesinas: una ecología política de transformaciones territoriales*, ed. Anthony Bebbington (Lima: IEP, 2011), 25, 28.
40. Estado Plurinacional de Bolivia, *Constitución Política del Estado* (2009).
41. Gaceta Oficial del Estado Plurinacional de Bolivia, *Ley no 3760 de7 de noviembre de 2007* (2007).
42. International Labour Organisation, Indigenous and Tribal Peoples Convention (No. 169) (1989).
43. United Nations Declaration on the Rights of Indigenous Peoples.
44. Ibid.
45. Lennox, 'Natural Resource Development', 18; Doyle and Gilbert, 'Indigenous Peoples and Globalization', 250.
46. Shin Imai, 'Indigenous Self-Determination and the State', in *Indigenous Peoples and the Law*, ed. Benjamin J. Richardson, Shin Imai, and Kent McNeil (Oxford: Hart Publishing, 2009), 292.
47. *Western Sahara*, Advisory Opinion 16 October 1975, (1975) ICJ Reports 12.
48. Margaret Satterthwaite and Deena Hurwitz, 'The Right of Indigenous Peoples to Meaningful Consent in Extractive Industry Projects', *Arizona Journal of International and Comparative Law* 22 (2005): 1–4.
49. Doyle and Gilbert, 'Indigenous Peoples and Globalization', 247–8.
50. Estado Plurinacional de Bolivia, *Constitución Political del Estado*.
51. Clayton Mendonça Cunha, 'The National Development Plan as a Political Economic Strategy in Evo Morales's Bolivia: Accomplishments and Limitations', *Latin American Perspectives* 37, no. 4 (2010): 177–96; Radosław Powęska, 'Powrót państwa wszechmocnego – ubóstwo i wykluczenie społeczne w Narodowym Planie Rozwoju Boliwii Evo Moralesa' [The Return of the Omnipotent State – Poverty and Social Exclusion in the National Development Plan of Evo Morales' Bolivia], in *Ubóstwo i wykluczenie. Wymiar ekonomiczny, społeczny i polityczny* [Poverty and Exclusion. Economic, Social and Political Dimensions], 193–201. (Warszawa: Bramasole, 2010); República de Bolivia, *Plan Nacional de Dessarrollo: Bolivia digna, soberana, productiva y democrática para Vivir Bien* (2006).
52. Elaboration by the author on the base of: Instituto Nacional de Estadística, http://www.ine.gob.bo
53. Ibid.

54. Georgina Jiménez, *Territorios indígenas y areas protegidas en la mira. La ampliación de la frontera de industrias extractivistas* (La Paz: Petropress, 2012), 7, 15, 17.
55. Gaceta Oficial de Bolivia, Decreto Supremo No. 0676 (2010).
56. Gaceta Oficial de Bolivia, Decreto Supremo No. 2366 (2015).
57. Carlos Corz, 'Bolivia llegaría a 48 TCF con reservas de shale gas y ocuparía 4to lugar en la región', La Razón, 23 August 2012, http://la-razon.com/economia/Bolivia-llegaria-TCF-reservas-ocuparia_0_1674432578.html (accessed 20 February 2015).
58. Mónica Oblitas Zamora, 'Fracking en Bolivia, la fractura de la madre tierra', *Los Tiempos*, 11 April 2015, http://www.lostiempos.com/oh/actualidad/actualidad/20150411/fracking-en-bolivia-la-fractura-de-la-madre-tierra_297906_657811.html (accessed 10 February 2016).
59. Mayorga, 'Bolivia: segundo gobierno', 24–5.
60. 'Ley de Consulta Previa no está en agenda del Legislativo', *Erbol*, 20 August 2015, http://www.erbol.com.bo/noticia/indigenas/20082015/ley_de_consulta_previa_no_esta_en_agenda_del_legislativo (accessed 15 January 2016).
61. Ministerio de gobierno, Estado Plurinacional de Bolivia, *Anteproyecto de Ley de Consulta Previa Libre e Informada* (2014).
62. Estado Plurinacional de Bolivia, *Ley de Minería y Metalurgia* (2014).
63. Penelope Anthias, 'Territorializing Resource Conflicts in "Post-Neoliberal" Bolivia: Hydrocarbon Development and Indigenous Land Titling in TCO Itika Guasu', in *New Political Spaces in Latin American Natural Resource Governance*, ed. Havard Haarstad (New York: Palgrave Macmillan, 2012), 129–53.
64. 'Guaraníes piden hace 19 años titulación de Takovo Mora', *Erbol*, 24 August 2015, http://www.erbol.com.bo/noticia/indigenas/24082015/guaranies_piden_hace_19_anos_titulacion_de_takovo_mora (accessed 15 January 2016); Luis Fernando Heredia, 'Derechos sobre papel mojado: el conflicto en Takovo Mora', *Los Tiempos*, 19 September 2015, http://www.lostiempos.com/lecturas/varios/varios/20150919/derechos-sobre-papel-mojado-el-conflicto-en-takovo_316086_700676.html (accessed 15 January 2016).
65. International Network for Economic, Social & Cultural Rights, *Case of the Mayagna (Sumo) Awas Tingni Community v. Nicaragua*, https://www.escr-net.org/docs/i/405047 (accessed 15 January 2016).
66. IACHR, 'Report No 40/04 Maya Indigenous Communities of the Toledo District Belize', 12 October 2004, http://www.cidh.org/annualrep/2004eng/belize.12053eng.htm (accessed 15 January 2016).
67. Almut Schilling-Vacaflor, 'Contestations over Indigenous Participation in Bolivia's Extractive Industry: Ideology, Practices, and Legal Norms', GIGA Working Paper No. 254 (2014); Almut Schilling-Vacaflor, 'Rethinking the Consultation-Conflict Link: Lessons from Bolivia's Gas Sector', GIGA Working Paper No. 237 (2013).
68. Ibid.
69. Ibid.
70. Anthony Bebbington, 'The New Extraction: Rewriting the Political Ecology in the Andes', NACLA Report on the Americas, September/October (2009): 12–40; Anthony Bebbington, *An Andean Avatar: Post-neoliberal and Neoliberal Strategies for Promoting Extractive Industries*, Brooks World Poverty Institute Working Paper 117 (2010); Radosław Powęska, *Indigenous Movements and Building the Plurinational State in Bolivia. Identity and Organisation in the Trajectory of the CSUTCB and CONAMAQ* (Warsaw: CESLA, 2013), 275, 278–9.
71. Shawn Regan, 'Unlocking the Wealth of Indian Nations: Overcoming Obstacles to Tribal Energy Development', PERC Policy Perspective No. 1 (2014).
72. CSUTCB, Nueva Constitucion Plurinacional. Propuesta política desde la visión de campesinos, indígenas y originarios (La Paz, 2006); CSUTCB, Plan Estratégico de Vida 2008–2017 (La Paz, 2008), 55.
73. CONAMAQ, *Propuesta: Constitución Política del Estado Plurinacional Qullasuyu-Bolivia* (Uru Uru Marka, 2007), 12, 24–5.
74. María Zegada, Yuri Tórrez, and Patricia Salinas, *En nombre de las autonomías: crisis estatal y procesos discursivos en Bolivia* (La Paz: PIEB, 2007), 65, 68–9.
75. Pacto de Unidad, Propuesta consensuada del Pacto de Unidad. Constitución Política del Estado Boliviano. Por un Estado Unitario Plurinacional Comunitario, Libre, Independiente, Soberano, Democrático y Social, (Sucre, 2007), 19, 21.

76. Álvaro Garcia Linera, 'El capitalismo andino-amazonico', *Le Monde Diplomatique* 123, January 2006.
77. Bebbington, '¿Una nueva extracción, una nueva ecología política?', 30–1.
78. David Harvey, 'The "New" Imperialism: Accumulation by Dispossession', *Socialist Register* 40 (2004): 63–87.
79. Peter B. Evans, 'Predatory, Developmental, and Other Apparatuses: A Comparative Political Economy Perspective on the Third World State', *Sociological Forum* 4, no. 4 (1989): 561–87, 562.
80. Roberto Acosta, 'Maldiciones que amenazan la democracia', *Nueva Sociedad* 229 (2010): 42–61.
81. Michael Ross, 'Does Oil Hinder Democracy', *World Politics* 53 (2001): 325–61.
82. Marco A. Gandarillas G, 'Bolivia: la década dorada del extractivismo', in *Extractivismo: Nuevos contextos de dominación y* resistencias, ed. Marco Gandarillas Gonzáles (Cochabamba: CEDIB, 2014), 103.
83. Jason Tockman and John Cameron, 'Indigenous Autonomy and the Contradiction of Plurinationalism in Bolivia', *Latin American Politics and Society* 56, no. 3 (2014): 57, 60–4.
84. Eduardo Gudynas, 'Estado compensador y nuevos extractivismos. Las ambivalencias del progresismo sudamericano', *Nueva Sociedad* 237 (2012): 128–46.
85. See James O'Connor, *The Fiscal Crisis of the State* (St Martins Press: New York, 1973).
86. Gudynas, 'Estado compensador', 137–8.
87. Salman, 'El estado, los movimientos sociales'; Mayorga, 'Bolivia: segundo gobierno'; Powęska, *Indigenous Movements*, 231–3; For the analysis of neo-developmentalism in Latin America, see: Antonio Araníbar and Benjamín Rodríguez, *América Latina, ¿del neoliberalismo al neodesarrollismo?* (Buenos Aires: PAPEP/PNUD, Siglo XXI Editores, 2013).
88. Powęska, *Indigenous Movements*, 131–281; see also Salman, 'El estado, los movimientos sociales'; Mayorga, 'Bolivia: segundo gobierno', 28.
89. Roberto Laserna, José M. Gordillo, and Jorge Komadina, *La trampa del rentismo ... y como salir de ella* (Fundación Milenio: La Paz, 2011), 39, 105–8.
90. Ciudadanía, Comunidad de Estudios Sociales y Acción Pública, *Cultura política de la democracia en Bolivia, 2012. Hacia la igualdad de oportunidades* (Cochabamba: Proyecto de Opinión Pública de América Latina (LAPOP), 2012), 21.
91. Gerardo Honty, 'Economías con pies de petróleo', *América Latina en movimiento online*, 15 September 2016, http://www.alainet.org/es/articulo/180273 (accessed 16 September 2016).

# Ethnic rights and the dilemma of extractive development in plurinational Bolivia

Rickard Lalander ⓘ

The Bolivian constitution of 2009 has been classified as one of the most progressive in the world regarding indigenous rights. The indigenous principles of *Suma Qamaña/Vivir Bien/ Good Living* on the harmonious relationship between humans and nature are established in the constitution. Nonetheless, these rights clash with the constitutionally recognised rights of the nation state to extract and commercialise natural resources (mainly hydrocarbons and mining) under the banner of redistributive justice, welfare reforms and the common good, in this study labelled the dilemma of extractive development. The article is based on ethnographic fieldwork and combines a political economy perspective on the extractive dilemma, while similarly examining the tensions between ethnically defined rights in relation to broader human rights in terms of values and norms related to welfare and conditions of living. The ethnic identity is multifaceted in Bolivia. Large segments of the indigenous population prefer to identify in class terms. The class-ethnicity tensions have altered throughout history, according to changing socio-economic, cultural and political settings. A central argument is that, during Evo Morales' presidency, class-based human rights in practice tend to be superior to the ethnically defined rights, as a reflection of the dilemma of extractive development.

## Introduction

[With this constitution] we have left the colonial, republican and neoliberal State in the past. We take on the historic challenge of collectively constructing a Unified Social State of Pluri-National Communitarian law, which includes and articulates the goal of advancing toward a democratic, productive, peace-loving and peaceful Bolivia, committed to the integral development and self-determination of the peoples.[1]

   The Bolivian constitution of 2009 is undoubtedly among the most radical in the world regarding the incorporation of international human rights criteria and the recognition of specific indigenous rights.[2] As expressed above in the fragment of the preamble to the constitution, Bolivia is no longer a republic but a plurinational state, which is a direct acknowledgment of the indigenous custom to organise according to distinct ethno-cultural identification within the same nation state. Additionally, the indigenous ethical-

This article was originally published with errors. This version has been amended. Please see Corrigendum (http://dx.doi.org/10.1080/13642987.2016.1215915).

philosophical conceptualisation of *Suma Qamaña/Vivir Bien* (live well) on the harmonious relationships among human beings and with nature/the environment has been established as the backbone of the constitution and national development policies. These innovative reforms have been applauded worldwide and enhanced the ethno-ecologist image of Bolivia and the government of Evo Morales Ayma.

A principal endeavour of the government since 2006 is the ambition to decolonise society, the state and the economy, which is also reflected in the constitution. Historically, the Bolivian political economy had excluded the indigenous population. Mining and extractive capitalism and imperialism based on exploitation of the indigenous peoples as labour force have characterised the Bolivian political economy since colonial times.[3] The 2009 constitution strengthened the position and role of the state in the economy, as a response to the discontent with neoliberal global capitalism.

The Morales government, which together with Venezuela and Ecuador has been in the forefront of what has been labeled twenty-first century socialism, has repeatedly emphasised that the state should achieve control of extractive industries in order to fund welfare policies and to achieve economic development. Several scholars have debated the advancements of Bolivia's (and Ecuador's) recent political orientation in terms of a post-neoliberal project or progressive neo-extractivism.[4]

Not only indigenous human rights were strengthened in the new constitution and secondary legislation in Evo Morales' time, but also broader social rights: the right to decent living conditions and well-being, instrumentalised through targeted social programmes and other policies aimed at redistributive justice, what is occasionally referred to as *class-defined rights* in this study.

Regarding the state control of vital industries – mainly hydrocarbons, agro-business and mining – the constitution declares the industrialisation and commercialisation of natural resources to be a key priority of the state, though taking into consideration the rights of indigenous peoples and provided that revenues should be directed at the common good (articles 319 and 355),[5] as will be further discussed in due course.

The aim of this article is to analytically examine the complex liaisons between ethnic and broader social (class-defined) rights amidst the Bolivian political economy of resource governance. The article consequently adds to the debates on contentious resource governance and the relationship, contradictions and tensions between class and ethnicity amid Bolivian identity politics and the question of indigeneity.[6] An essential claim is that class-based human rights in practice tend to outplay ethnically defined rights, due to an *extractive development dilemma*.

The dilemma of state authorities is, consequently, to be able to deliver welfare for all, which requires economic resources. With the public control of strategic industries, the redistribution of wealth through extraction can be achieved; that is, provision of class-defined rights. The rights of indigenous peoples and of the environment are affected in situations where natural resources are extracted in indigenous territories. This follows the logics that have characterised extractive projects around the globe, wherein economy almost always outplays ethnic and conservational rights. Shorter-term politico-electoral and economic realities tend to create social pressure that the state perceives as more urgent than the conservation of the environment and the protection of indigenous communities. In a previous study the metaphor of a *straitjacket* was used to discuss this unsolvable equation, referring to ethnic and environmental rights in relation to the ambitions of progressive governments to use extractive incomes for welfare reforms,[7] that is, more class-based concerns.

On the one hand, the complex identity politics of Bolivian indigeneity are examined. On the other hand, the extractive dilemma of the Bolivian government is problematised, particularly regarding the discourse and moral justification amidst the implementation of extractive politics on behalf of state authorities and how these discourses and justifications relate to the identitarian elements of class and/or ethnicity. The study likewise considers how these relationships are perceived at grassroots or intermediary socio-political levels, that is, by indigenous spokespersons.

While discussing Bolivian identity politics amidst recent extractive development projects, an additional identitarian element cannot be evaded; namely, the ecologist trait within indigenous identity. The environmental dimension is central in recent contentious politics around extractive projects. While ethnic/indigenous rights evidently are separated from the rights of nature/the environment, they frequently coincide in concrete extractive settings. Therefore, the environmental or ecologist trait of actors and interests will in the future at times be incorporated in the ethnic identity.

As highlighted in a previous article in this journal, international institutions and treaties, such as ILO (International Labor Organization) Convention 169 of 1989 and the UNDRIP (United Nations Declaration of the Rights of Indigenous Peoples) of 2007 may be powerful, albeit symbolic tools for indigenous struggle worldwide. Nonetheless, regarding article 3 of UNDRIP on the 'exercise of indigenous peoples' foundational right to self-determination', the fulfillment of these rights depends on specific domestic politics and laws.[8]

However, neither a progressive legal framework, nor the ratification of international human rights standards concerning indigenous rights can in practice guarantee that these rights are always prioritised. There are contradictions in the constitution as well as in the national development plan, creating clashes between ethnic rights and social welfare goals (class-defined rights). As elsewhere in the continent, the (*de jure*) rights on paper do not necessarily correspond to the (*de facto*) rights in practice.[9]

Methodologically, the study draws on ethnographic fieldwork undertaken in Bolivia between 2010 and 2015; hundreds of semi-structured interviews have been realised, although only a selection of these will be mentioned. The ethnographic material is complemented with critical reading of previous scholarly work and legal documents. For the sake of the argument it will be necessary to simplify matters a bit regarding the tensions between ethnicity and class among the actors. In order to see more clearly the particularities of each process or identitarian component, a relative simplification is accordingly essential.

After the contextualisation above, the layout of the text is as follows. First, a concise background of the Bolivian indigenous movement struggle is offered, mainly from 1952 onwards and particularly emphasising the politicisation of ethnic identity and the essential tensions between class and ethnicity. Thereafter, the *Vivir Bien* philosophy and the contradictory nature of the 2009 constitution are examined, particularly those articles that reflect the values and rights of class and ethnicity and the extractive rights of the state. The Law of Mother Earth and the ecologist traits of the indigenous peoples are then briefly scrutinised. Next, the most emblematic case of ethno-ecologist resistance against extractive developmentalist policies is analysed: the TIPNIS conflict. The TIPNIS case illustrates the contradiction between indigenous rights claims and state practices. Subsequently, before closing the article with a few pertinent conclusions, indigeneity and the extractive development dilemma are further problematised through the analysis of distinctive positions of involved actors.

## Bolivian indigeneity and politicisation of ethnicity in retrospective

In different parts of Latin America, most remarkably in Mexico and Peru, the *Indigenismo* ideology spread during the first half of the twentieth century and was a strategy to deal with the 'Indian problem'. Non-indigenous intellectuals, priests and politicians advocated for the improvement of the social conditions of the indigenous populations, albeit typically with a paternalistic and clientelistic style and as a component of nation building based on *mestizaje* (miscegenation). For instance, historian Marc Becker concludes that: '*Indigenismo* was always a construction of the dominant culture, particularly that of elite intellectual mestizos who used Indigenous issues to advance their own political agendas.'[10]

In Bolivia, *Indigenismo* was articulated already in the 1920s in the attempts to integrate indigenous and mestizo workers, accordingly realigning the relationship between the non-indigenous working class and the indigenous peoples.[11] The most persistent expression of Bolivian *Indigenismo*, however, came with the National Revolution of 1952 and the Agrarian Reform of 1953, directed by the MNR party (*Movimiento Nacionalista Revolucionario/* Revolutionary Nationalist Movement). Until then, the native populations in both the highlands and lowlands identified precisely as Indians (*Indios*), but were thereafter redefined as peasants/*campesinos*.

The 1952 revolution thus brought a cultural 'peasantification' of the indigenous communities and cultural practices, and the traditional sociopolitical and organisational structures were dissolved. The decree of the Agrarian Reform of 1953 also signified that all direct mentions of 'Indian peoples', Aymara, Quechua, etc. were erased.[12] In retrospect, Quechua intellectual and educator Leonel Cerruto exposes the following:

> 1952 is a rupture, because it wasn't the Indigenous movement that established the fracture or the individual land property. It came as a current of the Left, copied from Mexico. Of course, the *Indigenismo* [and its slogans of] land for the peasants, land for those who work it. So, the 'peasantification' (*campecinización*) begins at that moment, also for the indigenous Bolivians. However, they do not finish being neither Indigenous nor native. No. This problem persists until today.[13]

Already in the 1960s and 1970s radical indigenous intellectual organisations reacted to this cultural usurpation and homogenisation. Young urbanised indigenous students gathered around the Quechua-Aymara writer and intellectual Fausto Reinaga who wrote and spoke of the importance of the ethnic and cultural roots of the indigenous peoples. Reinaga rejected the 'peasantised' identity and endorsed a return to indigeneity. Jesuit anthropologist Xavier Albó summarises the meetings around Reinaga: 'We were reduced to peasants ... Let's go back to being Aymaras.'[14] These activities led to the formation of two parallel movements: the *Katarismo* and *Indianismo*.

Historian Pedro Portugal Mollinedo militated in the *Indianismo* movement and he recalls that whereas *Indianismo* accentuated the cultural, racial and national aspects of the indigenous peoples, *Katarismo* concentrated on economic and social class dimensions of struggle.[15] Both of these movements – at times merged and occasionally opposing each other – challenged the dominant peasant class-intensive rhetoric of the period and onwards originated in different political organisations and discourses. The *indianistas* organised early into their own political parties, in 1960 as PAN (*Partido Agrario Nacional*), later in 1968 as *Movimiento Nacional Tupak Katari* and in 1978 as *Movimiento Indio Tupak Katari*/MITKA. A leading Aymara indianista was Constantino Lima. In his words: 'As Indians they subjugated us, as Indians we will liberate ourselves.'[16]

The maximum emblematic historical figure of both *Indianismo* and *Katarismo* was Túpak Katari who, before being executed by the Spaniards, led an indigenous rebellion in the early 1780s together with his mistress Bartolina Sisa. The *Kataristas* were in comparison more successful in organised peasant-indigenous struggle with their *Movimiento*

*Revolucionario Tupak Katari*/MRTK party. In the 1990s, an armed *katarista* guerrilla movement operated: the Tupak Katari Guerrilla Army/*Ejército Guerrillero Tupak Katari*, which included prominent names such as Felipe Quispe (who later would become president of the indigenous-peasant union CSUTCB (*Confederación Sindical Única de Trabajadores Campesinos de Bolivia*)) and the current Bolivian vice-president Álvaro García Linera.[17]

The later Leftist political turn with Evo Morales as president is the outcome of decades of indigenous social movements' struggles against neoliberal politics, which dominated as the economic doctrine since 1985. Indigenous movement struggle is *per se* a proposal of systemic change and a rejection of global capitalism, imperialism and oppression along the lines of ethnicity and class.

A crossroads regarding the indigenous struggle occurred in 1990, when the lowland indigenous confederation CIDOB (Confederation of Bolivian Indigenous Peoples) organised a great *March for Territory and Dignity* from Trinidad in the lowlands to La Paz. The indigenous peoples had struggled for decades for deepened autonomy and dignity; since 1989 this was further triggered by ILO Convention 169. Also, highland indigenous organisations intensified pressure during the period and, despite the neoliberal context, the 1990s saw several legal recognitions of the ethnically defined grievances, such as the Law of Popular Participation/LPP of 1994 and the approval of Native Communal Lands/ TCOs (*Tierras Comunitarias de Origen*). The Bolivian ratification of ILO Convention 169 in 1991 was the outcome of the CIDOB march in 1990. However, the concrete implementation of Convention 169 and the more comprehensive right to prior consultations on hydrocarbon activities in Bolivia were deferred until 2007, that is, after the presidential inauguration of Evo Morales.[18]

The period until 2003 was characterised by anti-neoliberal protest activities – road blockings, street manifestations, strikes, etc. – and indigenous organisations strengthened their central role. In the awakening of the new millennium, the growing movement around Morales was at the epicentre of several resource conflicts, most decisively during the so-called 'water war' (protests against the privatisation of water) in Cochabamba in 2000 and the 'gas war' (against the export of gas to the United States via Chile) in El Alto-La Paz in 2003, which concluded with the resignation of neoliberal President Gonzalo Sánchez de Lozada.[19]

## Now the indigenous social movements are government

In December 2005, the leader of the coca-growing peasants, Evo Morales, was elected president of Bolivia, the first indigenous president ever of Latin America, representing the political party/movement MAS-IPSP (*Movimiento Al Socialismo – Instrumento Político por la Soberanía de los Pueblos/Movement Towards Socialism – Political Instrument for the Sovereignty of the Peoples*), which was a creation of the principal Bolivian indigenous-peasant social movements. The election of Morales can be viewed as the culmination of a protest cycle against exclusionary political and economic structures and also as recognition of previously disregarded citizens and collectives (mainly the indigenous and the poor).

Undoubtedly, a relative *dignification* of indigenous Bolivians has taken place during the Morales era. While racism and discrimination of indigenous citizens still exist as fundamental ills of Bolivian society, being indigenous is nonetheless not constantly an obstacle, compared to the *pre-Evo* period. Under the banner of decolonisation, a gradual social transformation has occurred, with improvements concerning ethnically defined recognition. A common joke during the first years of Morales' presidency was that before going to a job interview, the best thing to do was to first dress up in a poncho and/or other ethnic symbols.

For some scholars, a redefinition of peasant-indigenous citizenship and of the connotation of the Bolivian nation took place. This was achieved through the combined struggle of the indigenous-peasant base organisations and the political-electoral instrument (MAS-IPSP, onwards referred to as MAS) that, via the channels of representative democracy, succeeded in occupying the state.[20]

Some words are needed regarding the principal organisations of the indigenous movement. These have different profiles concerning class and/or ethnicity. The already mentioned lowland confederation, CIDOB, is traditionally associated with an ethnic profile, as reflected also in its name (Confederation of Bolivian Indigenous Peoples). The second, more indigenous of the organisations, or perhaps even native[21] (*originario*),[22] is the highland Aymara and Quechua Council CONAMAQ (*Consejo Nacional de Ayllus y Markas del Qullasuyu*). The remaining four organisations are in comparison more class-intensive; the *Bolivian Workers Central*/COB (*Central Obrera Boliviana*), the already mentioned peasant union CSUTCB, the *Bartolina Sisa National Confederation of Peasant, Indigenous, and Native Women of Bolivia*/CNMCIOB-BS (*Confederación Nacional de Mujeres Campesinas Indígenas Originarias de Bolivia Bartolina Sisa*), and the *Unionist Confederation of Intercultural Communities of Bolivia*/CSCIB (*Confederación Sindical de Comunidades Interculturales de Bolivia*).[23] The backbone of this last organisation – the CSCIB – is the coca-growing migrants, originally from the highlands, but for many years established also in the lowlands. These four organisations support the MAS government, whereas CONAMAQ and CIDOB have divided.[24] Nonetheless, all six organisations participated in the processes of rewriting the constitution.

Even though I argue that class and ethnicity are complexly merged in all organisations mentioned here – and that the differences frequently tend to be more semantic – it is interesting that some authors make a distinction between indigenous and native and classify CONAMAQ as native and CIDOB as indigenous. The *Bartolina Sisa* is labelled peasant/class-intensive, but includes both indigenous and native in its name.[25] The CSUTCB was born out of the *Katarista* movement, albeit backed up by the Catholic Church, NGOs (non-governmental organisations) and leftist political parties.[26] In one of its first official documents, the political thesis of the confederation in 1979, the organisational identity is expressed as 'Aymara, Quechua, Campa, Chapaco, Chiquitano, Moxo, Tupiguaraní and other peasants',[27] that is, both peasant and indigenous. According to its current organisational statutes (art.3.1), the CSUTCB is the:

> … maximal natural and historical organization that represents the totality of Indigenous-native-peasant nations and peoples of the Plurinational State, the setting of historical struggles of the indigenous-native-peasant movements in pursuit of the liberation of the colonial and republican state.[28]

As early as 1994, the CSUTCB agreed upon the strategy to form a 'political instrument', specifically to be able to 'create our own state, with our own constitution, in which indigenous nations can produce their own state' and this meeting can be considered the embryo of the route towards the constitution of 2009.[29]

## *Vivir Bien*, ethnic rights and the extractive dilemma in the constitution of 2009

The indigenous ethical-philosophical principles of *Suma Qamaña* (Aymara) – in Spanish translated to *Vivir Bien* (Live Well) – are incorporated as the backbone of the new constitution and the national development plans in Evo Morales' era. As often voiced by Morales and Bolivian Foreign Minister David Choquehuanca (who like Morales is Aymara), the idea is to live 'well' and not 'better' (at the cost of others), that is, not strive for material and economic accumulation, to increase 'living conditions' on a personal level, better than the neighbours, etc. Interestingly, the more ecologically concerned faction of the government – the *Pachamamistas* – can be traced back to a group within the *Indianismo* movement decades earlier. Incidentally, Choquehuanca has publically identified as *Indianista*.[30]

Advocates of Suma Qamaña suggest that development/progress as most people understand it is unnecessary. Individual as well as national economic 'progress' and 'wellbeing' regarding material belongings and capital accumulation and so forth, according to traditional notions of development, should be compared with a life in harmony with the environment and other human beings, without the burdens of global capitalism on nations and of consumptionism on individuals and collectives. Academics, social movement activists and politicians in Bolivia, Ecuador and elsewhere frequently use the *Vivir Bien* concept, both as a critique of development (understood as progress/economic growth) and as a principle of harmonious and ecologically sustainable life.[31]

Now, for a better comprehension of the legal setting and the complexities amidst the dilemma of extractive development, the ethnic-indigenous as well as broader social rights, and also the 'extractive developmentalist' rights of the state, it is of great importance to examine some crucial parts of the 2009 constitution.

In its preamble, the constitution declares that Bolivia is no longer a republic but a plurinational state. The plurinational character of the 'refounded' Bolivia embraces several dimensions. The first is related to decolonisation and the historical-symbolic recognition of legal pluralism, indigenous autonomies in parallel to the traditional politico-territorial division of the state, as well as a broader reconfiguration of the political society based on indigenous participation. Another element is precisely that of the plurality of nations, which promotes the rights of different indigenous nations to articulate political demands and administer justice within the Bolivian nation.[32]

A novelty of the constitutional text relates to the central identitarian concepts of this study. In each citing of the ethnically defined peoples, indigenous is grouped together with peasants/*campesinos* and natives/*originarios*: *indígena-campesino-originario*, that is, as indigenous-peasant-native peoples. This innovation will be discussed later.

The indigenous identity and specific indigenous rights are specified in articles 30–32; for instance the right to self-determination and territoriality. Regarding who is to be classified as indigenous, the first part of article 30 declares:

[They are] the rural native indigenous people and nationality consisting of every human collective that shares a cultural identity, language, historic tradition, institutions, territory and world view, whose existence predates the Spanish colonial invasion.

In the same article, the rights to prior consultation and *free, prior, informed consent/FPIC* of the indigenous peoples are mentioned. The right to prior consultation and FPIC of the affected indigenous peoples before extractive activities take place in their territories is a crucial component of the dilemma of extractive development. Even if this legal and rights-based mechanism has been practiced more frequently in Bolivia than in the majority of neighbouring Latin American countries, numerous shortcomings have characterised these procedures, such as insufficient decision-power and information of the involved population, irregularities and lack of transparency as concerns compensation payments.[33]

Another important recognition of indigenous rights is that of *indigenous native peasant autonomy/Autonomía Indígena Originaria Campesina* (articles 289–296), likewise accompanied by secondary legislation in 2009. Notwithstanding, as argued by Canessa, the government resisted the ambitions of numerous indigenous communities that tried to achieve the status of autonomous indigenous municipalities via a referendum. The argument of the government was principally that 'the state was already indigenous' and the unity of the indigenous peoples was thus at stake should they to go ahead with the procedure.[34]

The protection of natural parks and indigenous territories is pronounced in article 385, whereas, the territorial rights of indigenous peoples are expressed in articles 394–395:

> The State recognizes, protects and guarantees communitarian or collective property, which includes rural native Indigenous territory, native, intercultural communities and rural communities. Collective property is indivisible, may not be subject to prescription or attachment, is inalienable and irreversible, and it is not subject to agrarian property taxes. Communities can be owners, recognizing the complementary character of collective and individual rights, respecting the territorial unity in common.

Nonetheless, in parallel to the collective indigenous rights, numerous individual and universal rights are expressed (some of which evidently are rooted in indigenous culture), such as in article 8:

> The State is based on the values of unity, equality, inclusion, dignity, liberty, solidarity, reciprocity, respect, interdependence, harmony, transparency, equilibrium, equality of opportunity, social and gender equality in participation, common welfare, responsibility, social justice, distribution and redistribution of the social wealth and assets for wellbeing.

Broadly speaking, throughout the constitution there are references to the central objectives of poverty reduction, welfare provision, economic development and environmental protection (e.g. article 312). Moreover, articles 306 and 313 emphasise that the overarching ambition of Bolivian economic policies is to overcome poverty and social/economic exclusion [class-defined rights]; consequently the following sub-goals are identified:

1. The generation of social wealth within the framework of respect for individual rights, as well as the rights of the peoples and nations.
2. The fair production, distribution and redistribution of wealth and economic surplus.
3. The reduction of inequality of access to productive resources.

However, as mentioned in the introduction, the same constitution equally expresses the rights of the state to explore the natural resources of the soil, as pronounced in articles 319 and (below) 355, which also indicates the destination of the incomes derived from these activities:

I. The industrialization and sale of natural resources shall be a priority of the State.
II. The profits obtained from the exploitation and sale of the natural resources shall be distributed and reinvested to promote economic diversification in the different territorial levels of the State. The law shall approve the percentage of profits to be distributed.
III. The processes of industrialization shall be carried out with preference given to the place of origin of the production, and conditions shall be created which favor competitiveness in the internal and international market.

Clearly, prevailing economic and political interests conflict with indigenous-territorial and environmental rights. This enigma – reinforced rights and the maintenance of resource extraction reliance – is clearly expressed in the constitution. National authorities justify the persistent extraction with the necessity to achieve distributive justice, that is, a diminution of poverty and the provision of welfare for all, especially the marginalised sectors. This approach, with the partial sacrifice of the specific rights of the environment/nature and indigenous peoples to achieve social welfare is sometimes labelled progressive neo-extractivism.[35]

### The Law of Mother Earth and the ecologist traits of the indigenous peoples

Reconnecting to *Vivir Bien* and the ethno-environmental profile of Bolivia, in 2010 the *Law of Mother Earth* was established. The law embraces, among others, the rights to the protection of the integrity of life and natural processes, the continuation of vital life cycles and processes free from human alteration and to not be affected by mega-infrastructure and development projects that disturb the balance of ecosystems and local communities. In 2012 the law was upgraded by the National Legislative Assembly as the *Framework Law of Mother Earth and Integral Development to Live Well/ Ley Marco de la Madre Tierra y Desarrollo Integral para Vivir Bien.*[36] The framework character of the law implies that it is superior to other laws. For instance, the laws of mining, hydrocarbons, water, etc. should correspond to the contents of the framework law. The second part of the law should be emphasised, namely the *integral development* which refers to the objectives of *Vivir Bien* beyond merely environmental concerns and the proper rights of nature. Even if the environmental concerns are fundamental, the integral development component adds a more pragmatic dimension in considering human needs and rights as well.[37]

Nicole Fabricant reasonably criticises the 'indigeneity' component of the discourses based on ecologism and the sustainable ways of living by the indigenous population, as these are based on rural populations, with a close relationship to the territory. She therefore questions the relevance of these logics for the large urban indigenous population. Likewise, Fabricant is sceptical vis-à-vis the *Vivir Bien* alternative provided to confront the climate crisis.[38]

A few words on the possible ecologist trait of the indigenous peoples are required. The notion of the 'ecologically noble native' has been used in discourses both by the Morales government and by oppositional indigenous activists. However, whether this identitarian feature of the indigenous is natural, intrinsic or automatic might be questioned. For some scholars, the 'ecological Indian' is one of the oldest myths of anthropology[39]; or it might be understood as partly constructed identities, that is, through the strategic fusion of ecologist and indigenous discourses at local and transnational levels. As Astrid Ulloa convincingly argues, indigenous organisations have been successful in being perceived around the globe as the prime defenders of the interests of Mother Earth, as frequently is observable at international climate summits.[40]

This is not to claim that the environmental concern of the indigenous organisations should not be taken as a serious grievance. However, one should not take completely for granted that this ecologist concern is automatic. Indigenous peoples are not a homogenous group and other identity-based concerns might be superior in specific situations, such as questions of poverty and ethnic discrimination.

Indigenous intellectual NGO activists Edwin Armata Balcazar and Walter Limache Orellana clarify that ecologism may not generally constitute the superior element of indigenous identity, but concerned indigenous peoples are inclined to react when extractive projects threaten their traditional livelihoods.[41]

## The TIPNIS conflict

The national park and indigenous territory of TIPNIS (*Territorio Indígena y Parque Nacional Isiboro Secure*) is the home of, among others, the Mojeño-Ignaciano, Yuracaré and Chimán indigenous peoples and is similarly a territory of rich biodiversity. Since 2011, TIPNIS has been the most emblematic Bolivian case of contentious politics between the state and indigenous and environmental activists. Between 15 August and 19 October 2011, a huge number of indigenous and other activists marched for 65 days from the Bolivian lowlands to the capital protesting against a highway construction project through the protected area and indigenous territory.

The population living in TIPNIS is polarised regarding the highway. There are groups of people in TIPNIS that support the highway, such as the migrating coca workers that see economic opportunities with the improved infrastructure. Throughout the conflict, the government justified the TIPNIS project with arguments related to social welfare, such as the promotion of health, education and infrastructure for isolated indigenous groups.

The 602-kilometre highway project would connect the lowland Beni department with highland Cochabamba and was a key project of Latin American integration by the new progressive governments following the commodity export strategy. Brazil was deeply involved in supporting the road so as to facilitate access to Pacific markets. The Brazilian Development Bank/BNDES financed the highway project and the Brazilian construction company OAS was contracted, which together with the interests of oil company Petrobras underscore the shifting continental geopolitical order. The Initiative for the Integration of the Regional Infrastructure of South America (IIRSA) is a strategic project of regional integration for the common good (welfare reforms), which is the technical project of the South American Council of Infrastructure and Planning (COSIPLAN) of UNASUR (Union of South American Nations). However, Vice-President Álvaro García Linera categorically denounces the alleged relationship between IIRSA and the TIPNIS highway.[42]

The protesters – led by the territorial organisation *Subcentral TIPNIS* and the CIDOB and CONAMAQ indigenous confederations – presented a list of 16 demands concerning respect for the territory as well as other social, economic and cultural concerns. After a few violent clashes between police forces and the marchers, President Morales agreed to all demands presented. Nonetheless, after a few months the highway project was re-initiated, despite heavy resistance and international media and academic coverage.[43]

Celso Padilla is leader of the Assembly of the Guaraní People/APG. During the march he declared that:

> We want to tell the Government that 'this' is the Plurinational State. Here we are the 34 indigenous groups. We want to tell the President Evo Morales that the State is built with these people. He should not concentrate the power only on CSUTCB, the 'Bartolinas' and the intercultural communities. We, these indigenous groups took him to power. Why he forgets so quickly?[44]

It is revealing that Padilla directly refers to the 'factionalism' of the indigenous movements and mentions the three most devoted supporters of the Morales government (CSUTCB, the Bartolinas and the Interculturals, that is, the more class-intensive organisations, in comparison with the ethnic profile of CIDOB to which the APG is associated).

The TIPNIS conflict unexpectedly ended – at least momentarily – in a rather peculiar way. During fieldwork in Bolivia in January 2014, the local academics' and activists' surprise was apparent regarding announcements (as quoted in local mass media on 4 January) by Vice-President García Linera vis-à-vis the destiny of the TIPNIS highway in June 2013 at a conference in Argentina – seven months earlier. García Linera acknowledged several mistakes committed by the government concerning the construction project and communication with affected indigenous groups (among others regarding the right to free, prior, informed consultation). He stressed that the highway will be indispensable but that it had to be postponed 20, 50 or 100 years and should be cautiously realised with regard to environmental protection.[45] During interviews in La Paz in January and February 2014, several government spokespersons confirmed that the highway construction was cancelled 'because the vice-president said so'.

Yet, this position of the government was rapidly altered after the presidential elections of October 2014, when Morales was re-elected by a landslide. So, the government decided once again to move onwards with the TIPNIS highway. Moreover, on 20 May 2015, Morales drafted and established a presidential decree that profoundly altered the rules of the political game in the protected areas, some of which are also indigenous territories. Morales reasoned that the national parks had been established by elite groups in Bolivia and beyond as a kind of standby supply of natural resources. State authorities plan to initiate oil drilling in eight of the existing 22 protected areas of the nation.[46] According to superior decree 2.366, the hydrocarbon resources in all parts of the nation could be extracted and commercialised for the common good (social welfare/poverty reduction).[47]

Moreover, Vice-President García Linera declared that a legal amendment of the statute on the intangibility of TIPNIS should be realised,[48] which was one of the central accomplishments of the 2011 indigenous march for TIPNIS. Based on public statistics and according to calculations by CEDIB (Bolivia's Documentation and Information Centre) oil and gas concessions today overlap 11 of the 22 Bolivian protected areas. In TIPNIS, 35% of the territory is assumed to be open for oil and gas extraction according to these figures.[49] Morales presents other statistics concerning the proportions affected by extractivism. According to the president, the extractivism will affect only 0.0008% (205 hectares) of the territory of seven protected areas (in total 3.9 million hectares, each area with an average size of 560,000 hectares). Regarding the right of the affected population to prior, free and informed consultation, heavily debated throughout the TIPNIS conflicts, Morales has repeatedly expressed that he considers this procedure to be a waste of time and money, but that this right of the peoples should be respected.[50]

In early October 2015, the Ministry of Energy and Hydrocarbons publically announced that it had the support of around 50 affected indigenous communities (mostly Guaraníes) in the departments of Tarija, Santa Cruz and Chuquisaca, concerning the exploitation of hydrocarbons in the protected areas. The Minister of Energy and Hydrocarbons, Luis Alberto Sánchez, referred to letters from the concerned indigenous groups, which expressed their backing of development plans in their regions. As they argued, they wished to 'have the same opportunities as elsewhere as regards, schools, hospitals and roads'.[51]

Bolivia's Minister of Planning, sociologist René Orellana, reflects on the TIPNIS conflict and existing ethical and economic challenges:

> Bolivia needs that highway. And we need to take advantage of the existing hydrocarbon resources. Why not? We have to do it, although respecting the limits of regeneration and searching for the best technological conditions. So: why not? Sometimes I don't understand how certain actors say: 'We can't do that; the rest of Bolivia should not be benefitted by those resources.' That's an egoistic position. We need the complementarity of rights. All rights are exclusive; the economic, social and cultural rights, the rights of the poor and of all citizens are superior.[52]

Summing up, as reflected above, the TIPNIS case can be viewed as an illustration of the superiority of class-defined rights vis-à-vis ethnic rights, for instance when the justification of the government is exposed in discourses.

### Indigeneity and the dilemma of extractive development

> In almost ten years of the cultural and democratic revolution, Bolivia has positioned itself as an international reference. Initiated in the social movements, our own economic, social and communitarian model is independently established. We have recovered our natural resources that will allow us to redistribute the revenues, benefiting all Bolivians (Evo Morales Ayma, Public Discourse, August 2015).[53]

The capitalist logics of accumulation are still central traits of the Bolivian political economy, which has been criticised by many activists and scholars that were hoping to witness the progress of an anti-capitalist/post-capitalist project in the country.[54] However, the Morales administration has since the beginning explicitly communicated that the state should attain control of extractive industries so as to finance welfare reforms and to achieve economic development. Moreover, as argued by Almut Schilling-Vacaflor, the Morales government realised radical legal reforms regarding both human rights and environmental principles within the hydrocarbon sector. These improvements, including the acknowledgment of rights in the 2009 constitution, were the outcomes of decades of popular struggle, principally by lowland indigenous peoples.[55]

Evidently, the debates on indigeneity and the rights of indigenous peoples as expressed, for instance, by NGOs and the United Nations, might function better in contexts of nation states in which the indigenous peoples are a minority. In Bolivia – since 2006 sometimes categorised as an 'indigenous state' – the indigenous peoples constitute the majority. The Bolivian case displays a series of complexities and contradictions regarding claims of indigenous rights and the meaning of indigeneity.[56] By the same token, the indigeneity discourse has strategically been used both by government and oppositional activists.[57]

During interviews in the Amazon (Beni department), I was privileged to discuss the territoriality element as the fundamental criteria of indigeneity (referred to in article 30 of the constitution) with a Mosetén leader and a non-indigenous activist. Both expressed a rather fundamentalist definition of territoriality as a superior criterion of indigeneity, and, following that logic, neither urban nor migrating Bolivians of indigenous heritage would qualify as native-indigenous. For the same reason they argued that the only 'pure' native-indigenous organisations would be CONAMAQ and CIDOB.[58]

Quechua educator Leonel Cerruto does not agree. He claims that also urban and migrating indigenous individuals and collectives fulfil the criteria of indigeneity, as also the CSUTCB, the Bartolinas and the intercultural organisations. In the first place, he argues, the indigenous peoples have always been mobile; therefore it is frequently complicated to define the precise native origin of the ethnic group. Moreover, in reference to the constitution and the internal quandaries of the broader indigenous movement, he reflects:

> Indigenous-native-peasant without commas enters the constitution as one word in order not to cause fragmentation or division between us, since indigenous, natives as well as peasants deal with territoriality. Our principle has always been land *and* territory, i.e. both things: land *and* territory as principles of life. In that way we have been able to rearticulate, but still we have this false vision of the peasant not being indigenous. The only thing that brings us is division, when we really need to unify.[59]

In a previous conversation of early 2011, the CSUTCB leader Damián Condori articulated the following regarding identification in terms of class and/or ethnicity (in relation to the formulations in the new constitution):

> Yes, we are generally getting used to refer to ourselves as indigenous-natives-peasants. That is, I am not just a peasant any longer. We are peasants, indigenous and natives. Well, we are the majority. Almost 65% of the population identify as indigenous, natives and/or peasants. And peasant is the most common identity in Bolivia, and this is due to the presence of the CSUTCB in the nine departments. We have executives in the nine departmental federations, whereas the CONAMAQ – who identifies as natives – and was founded only in 1998, is merely marginally important in some places, such as Oruro, parts of La Paz and Potosí.[60]

Anders Burman provides an interesting observation on fluctuating ethnic identities. Earlier highland indigenous peoples identified as 'natives' and referred to the lowland peoples as indigenous. Burman recalls his meetings with CONAMAQ authorities, that in 2001 were perplexed when being labelled 'indigenous/*indígena*', since the indigenous – according to them – were the lowland peoples, whereas the highland Aymara and Quechua identified as native. In 2011, during the TIPNIS protests, this cultural-semantic identification has altered and CONAMAQ spokespersons now identified and marched as 'indigenous', that is, identifying with a broader struggle, involving also lowland indigenous peoples.[61] The strategic indigeneity may thus evolve in different directions, either through the 'ethnification' of previously class-intensive actors, or, as this case with CONAMAQ, going from native to indigenous.

The indigenous *and* class-defined discourse of Evo Morales and his government is pronounced and directed at different levels: the domestic and global spheres respectively. Being of Aymara origin, the ethnic identity of the president is extremely important in terms of symbolic ethnic capital. Even though he is first and foremost a peasant unionist and leader of the coca growers, his indigenous identity, albeit merged with classism, is essential in both domestic and internationally targeted speeches.

Nonetheless, regarding the reinforced position of indigeneity in the updated form of citizenship – particularly in comparison with the earlier recognition of indigenous peoples as peasants as a result of the 1952 Agrarian revolution – it is indispensable to provide a broader picture when examining the indigeneity within the political project of the Morales administration. Evo Morales has indeed been portrayed as a climate hero around the world, leaning on discourses based on indigenous values and the worldview of *Vivir Bien (Suma Qamaña)* as options for responding to both global capitalism and the climate crisis. But, this discourse is applied mostly at a global level, whereas the domestic speeches of Morales deal more with development economics and fair distribution of resources, that is, policies and rights defined by class and social justice.[62]

Jorge Viaña Uzieda is currently the head of a government institution on the issue of indigenous autonomy. Regarding the challenges of political, social, economic and ecological transformation, he espouses the following:

> I do not think that Bolivia will resolve the dilemma between development and preservation. That is, the priority of what has been labeled development until now, or less devastating alternatives to development, has been opened and suggested, although as a dilemma, as a question to be resolved … They have to try to harmonize the relationship between certain necessity to defend and safeguard the environment and the pendant undeniable assignment to create conditions of welfare for the population. And, well, that implies extractivism. Let's be frank, from where will we take the resources? You have to generate surplus and redistribute it. So, the dilemma is planted.[63]

The aim was consequently neither to abandon the matrix of capitalist development, nor to entirely end the pollution of nature through extractivism or to always respect the indigenous territories, but to establish the dilemma and propose the *Vivir Bien* as an alternative to the world. The relative superiority of welfare policies vis-à-vis environmental conservation and (indirectly) indigenous territorial rights is similarly expressed in the quotation.

The government generally applies a pragmatic approach towards ethnic and environmental rights. The 2009 constitution includes articles that favour such an attitude towards these rights, if revenues of extractive activities are used for the common good and economic redistribution.[64] Consequently, following this reasoning, if class interests define the prioritised politics aimed at the common good and economic redistribution, the ethnic and environmental rights may be sidestepped.

However, the relationships and tensions between class, ethnicity and ecologism are, as stated, more complex. Anthropologist John-Andrew McNeisch emphasises the complexity of Bolivian identity politics and indigeneity. As he argues, a frequent misunderstanding by analysts is found in the simplification and generalisation of the relationship between class and ethnicity. Neither should the TIPNIS conflict be interpreted as such, that is, as a battle between indigenous and environmental activists on the one hand and the developmentalist and class-defined position of the government on the other. Even if these specific interests indeed exist, the realities of ethnic and class-based identification of the indigenous peoples are more complex and in flux. The indigenous communities have adapted to capitalist, consumptionist and developmentalist practices for many decades, which have shaped outcomes regarding collective and individual identities.[65]

It should similarly be emphasised that the grievances of the rural indigenous populations towards the state – both among oppositional and government supporters – generally have more to do with infrastructure, health and education, than with ecological concern or ethnic-cultural rights, as perceived during recent fieldwork in the Amazon (Beni department), as well as in the lowlands of El Chaco (Santa Cruz department).[66]

Rounding off, the extractive dilemma has been characterised by recent years of contentious politics and resource governance in Bolivia. Often these conflicts have been portrayed as choices between ethnic and environmental values and rights on the one hand and – agreeing with McNeisch – economic, developmentalist and class-defined interests and values on the other. The realpolitik is always a question of choices and priorities, and there will always be a certain degree of compromise and sacrifice of specific rights, interests and values.

**Concluding remarks**

In this article I have examined the ethnically defined rights in relation to broader and class-defined human rights in the historical context of the extractive development politics of the Evo Morales government. On the one hand, ethnic rights (and identity) were scrutinised in relation to rights/identities defined by class, albeit bearing in mind that these identitarian bases are complexly and intimately intertwined in Bolivian contemporary history. On the other hand, the relative recognition of these rights was inspected from the angle of Bolivian political economy and conditioned by the dilemma of extractive development.

The ethnic identity of the Bolivian indigenous populations is multifaceted and a large segment of these have preferred to identify primarily in class terms, as peasants, although at the same time recognising their ethnic identity. Indigeneity in Bolivia is to a large extent conditioned by the element of class identification, particularly in the central organisations behind the Morales government. Bolivia is an exemplary case in showing that neither identities nor social movements can be seen as static categories. They change with historical and

social alterations and the relative positions of individuals and collectives in a specific structure. Broadly speaking, ethnic/indigenous identity has gradually been reinforced during the
Morales administration and it is possible to speak of this process in terms of decolonisation
and a relative dignification of the indigenous peoples.

The incorporation of the indigenous philosophy of *Vivir Bien* in the constitution and
national development policies has reinforced the ethno-ecologist profile of the Morales government, particularly at a global level. Likewise, as has been discussed, the government
uses the indigeneity and ethno-ecologist discourses strategically.

It has been argued that in the Bolivian practice, ethnic rights (intertwined with environmental rights) frequently tend to be downgraded in relation to the broader class-defined
rights as an outcome of the extractive dilemma. While indigenous rights were decisively
reinforced with the constitutional reform of 2009, these rights clash with the constitutionally recognised rights of the nation state to extract and commercialise natural resources
(mainly hydrocarbons and mining) under the banner of redistributive justice, welfare
reforms and the common good – class-defined rights. This is a crucial expression of the
dilemma of extractive development.

The relative superiority of class-defined rights vis-à-vis ethno-territorial rights is likewise apparent when examining the government's justification of extractive projects and
the contentious construction of the highway through the TIPNIS territory. Summing up,
on basis of the material examined in this study, it is possible to conclude that ethnic (and
environmental) rights are frequently being subordinated to broader class-defined social
rights, as a consequence of the dilemma of extractive development.

## Acknowledgements

The author wishes to express gratitude to all colleagues and informants in Bolivia during fieldwork,
particularly Elizabeth Huanca, Oscar Vega, Fernando Galindo and Xavier Albó, also the anonymous
peer-reviewer of the journal and René Kuppe for inspiring comments.

## Funding

This work was supported by Svenska Forskningsrådet Formas [grant award 2012–1828] as part of the
project 'Rights of Nature – Nature of Rights: Neo-constitutionalism and Ethno-ecologist Resistance in
Bolivia and Ecuador' for the period 2013–2016.

## ORCID

*Rickard Lalander* ⏺ http://orcid.org/0000-0002-2581-2588

## Notes

1. Estado Plurinacional de Bolivia, *Constitución Política del Estado* (La Paz: Estado Plurinacional de Bolivia, 2009), Preamble.
2. The constitutional reform of Bolivia was strongly influenced by the 1989 ILO (International Labor Organization) Convention 169 on the rights of the indigenous peoples and further inspired by the United Nations Declaration of the Rights of Indigenous Peoples (2007).
3. Henry Veltmeyer, 'Bolivia: Between Voluntarist Developmentalism and Pragmatic Extractivism', in *The New Extractivism. A Post-Neoliberal Development Model or Imperialism of the Twenty-First Century?* Henry Veltmeyer and James Petras, eds (London and New York: Zed Books, 2014), 80–113, quote at 83.
4. See, for instance, Arturo Escobar, 'Latin America at a Crossroads. Alternative Modernizations, or Post-Development?' *Cultural Studies* 24, no. 1 (2010): 1–65; Kepa Artaraz, *Bolivia. Refounding the Nation* (London: Pluto Press, 2012).
5. Estado Plurinacional de Bolivia, *Constitución Política del Estado.*
6. Nancy G. Postero, *Now We Are Citizens. Indigenous Politics in Postmulticultural Bolivia* (Stanford, CA: Stanford University Press, 2007); John-Andrew McNeish, 'Extraction, Protest and Indigeneity in Bolivia: The TIPNIS Effect', *Latin American and Caribbean Ethnic Studies* 8, no. 2 (2013) 221–42; Lorenza Belinda Fontana, 'Indigenous Peoples vs Peasant Unions: Land Conflicts and Rural Movements in Plurinational Bolivia', *The Journal of Peasant Studies* 41, no. 3 (2014): 297–319.
7. Rickard Lalander, 'Rights of Nature and the Indigenous Peoples in Bolivia and Ecuador: A Straitjacket for Progressive Development Politics?' *Iberoamerican Journal of Development Studies* 3, no. 2 (2014): 148–72.
8. Irène Belliera and Martin Préaud, 'Emerging Issues in Indigenous Rights: Transformative Effects of the Recognition of Indigenous Peoples', *International Journal of Human Rights* 16, no. 3 (2012): 474–88.
9. Markus Kröger and Rickard Lalander, 'Ethno-Territorial Rights and the Resource Extraction Boom in Latin America: Do Constitutions Matter?' *Third World Quarterly* 37, no. 4 (2016): 682–702.
10. Marc Becker, '*Indigenismo* and Indian Movements in Twentieth-Century Ecuador' (paper presented at the Congress of LASA (Latin American Studies Association), Washington, 1995).
11. George Gray Molina, 'Ethnic Politics in Bolivia: "Harmony of Inequalities" 1900–2000' (Working paper, Centre for Research on Inequality, Human Security and Ethnicity/CRISE, University of Oxford, 2007).
12. Ibid., 6.
13. Leonel Cerruto, Interview, El Alto, 23 November 2015.
14. Xavier Albó, *Movimientos y poder indígena en Bolivia, Ecuador y Perú* (La Paz: CIPCA, 2009), 36.
15. Pedro Portugal Mollinedo, Internet Interview, 10 February 2016.
16. Deborah Yashar, *Contesting Citizenship in Latin America. The Rise of Indigenous Movements and the Postliberal Challenge* (Cambridge: Cambridge University Press, 2005), 168–9.
17. Albó, *Movimientos y poder indígena*, 36–40; Yashar, *Contesting Citizenship.*
18. Almut Schilling-Vacaflor, 'Prior Consultations in Plurinational Bolivia: Democracy, Rights and Real Life Experiences', *Latin American and Caribbean Ethnic Studies* 8, no. 2 (2013): 202–20, quote at 207.
19. For instance: Albó, *Movimientos y poder indígena*, 43–67.
20. See, for instance, Fernando Luis García Yapur, Luis Alberto García Orellana, and Marizol Soliz Romero, '*MAS legalmente, IPSP legítimamente'. Ciudadanía y devenir Estado de los campesinos indígenas en Bolivia* (La Paz: PIEB, PNUD, 2014).
21. I choose to use *native* throughout the text, and not the concepts of 'first peoples' or 'aboriginal' that are sometimes used while translating the Spanish ethnic concept of *originarios.*
22. See Fontana, 'Indigenous Peoples vs Peasant Unions', 305–7; García Yapur, García Orellana and Soliz Romero, '*MAS legalmente, IPSP legítimamente'.*
23. For example, García Yapur, García Orellana and Soliz Romero, '*MAS legalmente, IPSP legítimamente'.*
24. Since December 2013, CONAMAQ is divided with one faction supporting the Morales government and the other in opposition. In late 2010 CIDOB divided and presently there are two CIDOBs, one pro-government and the other oppositional.

25. Fontana, 'Indigenous Peoples vs Peasant Unions', 305–7.
26. Pedro Portugal Mollinedo, Internet Interview, 10 February 2016.
27. Albó, *Movimientos y poder indígena*, 39.
28. Confederación Sindical Única de Trabajadores Campesinos de Bolivia/CSUTCB, Estatuto Orgánico de la CSUTCB, Santa Cruz, 30 July 2010, http://comisionorganica-csutcb.blogspot. se/2012/09/estatuto-organico-de-la-csutcb-aprovado.html
29. Artaraz, *Refounding the Nation*, 45.
30. Karin Monasterios, Pablo Stefanoni, and Hervé do Alto, *Reinventando la nación en Bolivia: movimientos sociales, Estado y poscolonialidad* (La Paz: Plural editores, 2007), 28. However, Choquehuanca was not a militant of the historical *Indianismo* movement. Rather he was formed through non-governmental organisation (NGO) activism.
31. For instance: Eduardo Gudynas, 'Buen Vivir: Today's Tomorrow', *Development* 54, no. 4 (2011): 441–7; Kepa Artaraz and Melania Calestani, 'Suma Qamaña in Bolivia. Indigenous Understandings of Well-being and Their Contribution to a Post-Neoliberal Paradigm', *Latin American Perspectives* 42, no. 5 (2015): 216–33; Rickard Lalander, 'Entre el ecocentrismo y el pragmatismo ambiental: Consideraciones inductivas sobre desarrollo, extractivismo y los derechos de la naturaleza en Bolivia y Ecuador', *Revista Chilena de Derecho y Ciencia Política* 6, no. 1 (2015): 109–52.
32. Catherine Walsh, 'The Plurinational and Intercultural State: Decolonization and State Re-Founding in Ecuador', *Kult* 6 (2009): 65–84.
33. Schilling-Vacaflor, 'Prior Consultations in Plurinational Bolivia'.
34. Estado Plurinacional de Bolivia, *Constitución Política del Estado*, Chapter VII; Andrew Canessa, 'Conflict, Claim and Contradiction', *Critique of Anthropology* 34, no. 2 (2014): 153–73, quote at 167.
35. Eduardo Gudynas, *Extractivismos. Ecología, economía y política de un modo de entender el desarrollo y la Naturaleza* (Cochabamba: CLAES and CEDIB, 2015); Lalander, 'Entre el ecocentrismo y el pragmatismo ambiental'.
36. Estado Plurinacional de Bolivia, *Ley Marco de la Madre Tierra y Desarrollo Integral para Vivir Bien* (La Paz: Asamblea Legislativa del Estado Plurinacional de Bolivia, 2012).
37. Lalander, 'Entre el ecocentrismo y el pragmatismo ambiental', 126–7.
38. Nicole Fabricant, 'Good Living for Whom? Bolivia's Climate Justice Movement and the Limitations of Indigenous Cosmovisions', *Latin American and Caribbean Ethnic Studies* 8, no. 2 (2013): 160.
39. Shepard Krech Ill, *The Ecological Indian: Myth and History* (New York: W. W. Norton & Co., 1999).
40. Astrid Ulloa, *The Ecological Native. Indigenous Peoples Movements and Eco-Governmentality in Colombia* (New York: Routledge, 2005).
41. Edwin Armata Balcazar and Walter Limache Orellana, Group Interview, La Paz, 18 December 2015.
42. Álvaro García Linera, *Geopolítica de la Amazonía. Poder hacendal-patrimonial y acumulación capitalista* (La Paz, Vicepresidencia del Estado Plurinacional de Bolivia, 2012), 58–65.
43. See for instance: Fundación Tierra, *Marcha indígena por el TIPNIS*; McNeish, 'Extraction, Protest and Indigeneity'. The 2011 march was neither the first nor the last march in defence of TIPNIS. In 2012 a similar manifestation was realised, but did not succeed in attracting the participation and media coverage of the 2011 march.
44. Ibid. Quotation in Spanish originally published in Fundación Tierra, *Marcha indígena por el TIPNIS. La lucha en defensa de los territorios*, Comunicaciones (La Paz: El País S.A. 2012), 56.
45. Página Siete, 'El Vicepresidente descarta carretera por el TIPNIS', 4 January 2014, http://www. paginasiete.bo/nacional/2014/1/4/vicepresidente-descarta-carretera-tipnis-10441.html; Álvaro García Linera, *Los desafíos del proceso de cambio en Bolivia* (Conference at the Centro Cultural de Cooperación Floreal Gorini, Buenos Aires, 27 June 2013), http://www.centrocultural.coop/ videos/la-patria-grande-alvaro-garcia-linera-1-.html
46. Página Siete, 'Bolivia se sumó a la corriente de explotar áreas protegidas', 24 June 2015, http:// www.paginasiete.bo/nacional/2015/6/24/bolivia-sumo-corriente-explotar-areas-protegidas-60974.html
47. Evo Morales Ayma, *Decreto Supremo No 2366*, Presidential Decree, Estado Plurinacional de Bolivia, La Paz, 2015.

48. El Día, 'Gobierno busca anular la "intangibilidad" del Tipnis', 28 June 2015, http://eldia.com.bo/index.php?cat=1&pla=3&id_articulo=174956

49. David Hill, 'Bolivia Opens Up National Parks to Oil and Gas Firms', *The Guardian*, 5 June 2015,http://www.theguardian.com/environment/andes-to-the-amazon/2015/jun/05/bolivia-national-parks-oil-gas

50. Observatorio de Industrias Extractivas y Derechos Colectivos/OIEDC, 'Evo: En la consulta previa se pierde mucho tiempo', 13 July 2015, http://oiedc.blogspot.se/2015/07/evo-en-la-consulta-previa-se-pierde.html

51. Página Siete, 'Indígenas de 50 pueblos dieron aval para tareas de exploración', 3 October 2015, http://www.la-razon.com/economia/Gobierno-indigenas-pueblos-dieron-aval-tareas-exploracion_0_2355964414.html

52. René Orellana Halkyer, Interview, La Paz, 31 January 2014. Orellana holds a PhD in sociology and was earlier the main spokesperson of Bolivia in climate summits. He was also co-author of the Law of Mother Earth.

53. Evo Morales Ayma, *Discurso presidencial 639* (Ministerio de Comunicación, Estado Plurina-cional de Bolivia, La Paz, 7 August 2015, http://www.comunicacion.gob.bo/sites/default/files/media/discursos/Discurso%20Presidencial%2007-08-15.pdf

54. Rebecca Hollender, 'Capitalizing on Public Discourse in Bolivia – Evo Morales and Twenty-first Century Capitalism', *Consilience: The Journal of Sustainable Development* 15, no. 1 (2016): 50–76.

55. Schilling-Vacaflor, 'Prior Consultations in Plurinational Bolivia', 207.

56. Canessa, 'Conflict, Claim and Contradiction'.

57. McNeish, 'Extraction, Protest and Indigeneity'; Fabricant, 'Good Living for Whom?'.

58. Interview, Rurrenabaque, 16 November 2015.

59. Cerruto, Interview, El Alto, 23 November 2015.

60. Damian Condori, Interview, Sucre, 10 January 2011.

61. Anders Burman, '"Now We Are Indígenas": Hegemony and Indigeneity in the Bolivian Andes', *Latin American and Caribbean Ethnic Studies* 9, no. 3 (2014): 247–71.

62. Author's analysis of public discourses and observations during fieldwork (2010–2015). See also: Nancy Postero, 'Protecting Mother Earth in Bolivia: Discourse and Deeds in the Morales Administration', in *Amazonía. Environment and the Law in Amazonia: A Plurilateral Encounter*, James M. Cooper and Christine Hunefeldt, eds (Brighton: Sussex Academic Press, 2013), 78–93; Fabricant, 'Good Living for Whom?'

63. Jorge Viaña Uzieda, Interview, La Paz, 20 January 2014.

64. Lalander, 'Entre el ecocentrismo y el pragmatismo ambiental'.

65. McNeish, 'Extraction, Protest and Indegeneity', 238.

66. Author's observation and interviews, November–December 2015.

# The international human rights discourse as a strategic focus in socio-environmental conflicts: the case of hydro-electric dams in Brazil

Marieke Riethof ⓘⒹ

This article examines the mobilisation of human rights in campaigns against hydro-electric dams in Brazil. The symbolic and legal power of human rights has allowed activists to challenge official accounts of the impact of dams while deploying domestic and international legal frameworks. Although the politicisation of natural resources in Brazil has limited the effectiveness of anti-dam mobilisations, an appeal to the human rights agenda has translated into a powerful critique of the social impact of Brazil's development agenda, thereby making a moral and legal claim for justice.

## Introduction

In 2009 the National Indian Foundation (Fundação Nacional do Índio, FUNAI) reported on an expedition to the Amazonian state of Rôndonia where a major dam, the Jirau hydro-electric complex, was under construction. Detailing evidence of indigenous groups in voluntary isolation living close to the dam, the report also found increased deforestation, invasions of indigenous territories and cases of previously rare illnesses such as malaria and hepatitis. The report added that indigenous groups had fled the region following threats to their territory, construction noise and several loud explosions nearby.[1] Following on from this, in 2011 a labour inspection at the Jirau Dam encountered 38 workers in forced labour conditions, a situation exacerbated by health and safety problems and inadequate accommodation,[2] which in turn provided the impetus behind a large strike that took place at the Jirau Dam in early 2012. Frustrated by the conditions they experienced, workers – often migrants from other Brazilian states attracted by job prospects at dam construction sites – proceeded to set fire to buses, their lodgings and various communal areas. Reporting from helicopters circling around the dam site, television journalists labelled the fires and blockades as destruction and vandalism rather than representing the latter as a strategy of resistance that had finally forced the construction companies to negotiate.[3] During a strike on 6 April 2013 at the Belo Monte dam, workers also threatened to set their lodgings on fire while attempting to unite with other labourers at faraway construction locations, however the police[4] managed to obstruct their efforts. The government was more concerned about property and dam construction than human rights and responded to the strike – as well as regular occupations of the building sites by local communities – by employing the authorities to protect the construction sites, and in many cases used the latter to repress protests and occupations.

Dam construction projects have thus become associated with a wide array of human rights infringements, from forced labour and violent repression of protests to displacement and the destruction of the natural environment and people's livelihoods. Consequently, since the 1980s dam construction sites in Brazil have turned into significant sites of contestation, involving protests drawn from local communities, indigenous groups, environmental activists and workers.[5] Citing the often irreversible impact of hydro-electric dams on communities and the environment, these anti-dam campaigners have questioned the social and environmental sustainability of Brazil's ambition to expand hydro-electric power generation. This article begins by arguing that the economic, political and symbolic significance of natural resources for Latin America has created a politicised situation in which progressive governments have promoted the exploitation of natural resources while often ignoring the social and environmental costs. The article then turns to discuss the political significance of hydropower for Brazil's national development agenda, which has limited the space for opposing voices to be heard, despite Brazil's legally enshrined commitments to consult and protect affected groups. To circumvent these limitations, as the following section explains, anti-dam protestors have deployed a human rights agenda in order to frame socio-environmental conflicts in terms of international human rights, thereby highlighting the discrepancies between Brazil's ambitions for global leadership in environmental sustainability and human rights, and domestic realities. Finally, two cases of dam conflicts – the Dardanelos and Belo Monte dams – illustrate the political dynamic at work in socio-environmental conflicts in Brazil. At a domestic level, both conflicts show that anti-dam campaigners have strategically mobilised alternative accounts of the effects of dams on local communities, by simultaneously challenging official impact studies and widening the definition of the communities and territories affected. Both cases underline the significant role of the domestic legal system in challenging various aspects of the government decision-making and consultation process. However, with the government unresponsive to these demands due to the political and economic significance of hydropower, the Belo Monte campaigners have utilised an international human rights discourse to exert further pressure on the political process. The article concludes that while the international human rights framework has provided activists with a powerful political resource to mobilise for procedural rights to consultation and information, domestic and international legal strategies have proven to be a double-edged sword. The politicisation of natural resources in Brazil has meant that this strategy could not resolve substantive problems, such as the irreversible damage to local communities caused by dam construction.

## The politicisation of natural resources in Latin America

Conflicts about hydro-electric dams cannot be understood without reference to the political context in which they take place, as natural resources have become an increasingly problematic and politicised source of development. While reliance on natural resources is not a new phenomenon in Latin America, since the early 2000s the economic, political and symbolic significance of the sector has deepened at the global, national and local levels. David Harvey's[6] concept of accumulation by dispossession suggests how the global expansion of capitalism – with the state playing a significant role in this process in Latin America – has intensified the commodification of nature and environmental degradation. The creation of new development frontiers through the expansion of agriculture, infrastructure and resource extraction in regions such as the Amazon has become associated with the dispossession of communities living near mega development projects in terms of their land, culture and livelihoods. The contestation of natural resource exploitation has thus become inextricably linked with the development agenda pursued by Latin American governments, reflecting the political and symbolic significance of natural resources. The politicisation process therefore involves both the government discourse about the importance of natural resources for national economic and social development and the intensifying contestation surrounding the social, political and environmental sustainability of extractivist projects.

The politicisation of natural resources since the turn of the twenty-first century followed a period during which Latin American governments introduced neoliberal policies to reduce the role of the state, including political interference in the natural resources sector. In the aftermath of the 1982–1983 debt crisis and under pressure from global financial institutions, governments in Latin American countries such as Argentina, Chile, Brazil and Mexico decided to privatise key state-owned enterprises, including banks and companies in the electricity and natural resources sector. For example, the Brazilian government privatised the iron ore mining company Companhia Vale do Rio Doce in 1997, now the country's largest exporter. During the 1980s and 1990s we can therefore speak of the depoliticisation of natural resources[7] as their exploitation shifted away from the state to the private sector, with a focus on shareholder profits rather than national developmental concerns in an attempt to exclude political considerations from shaping the sector's future.

Since the late 1990s the Latin American political spectrum has changed significantly as voters elected left-wing parties, including in Bolivia, Brazil, Ecuador and Venezuela. What these countries have in common in addition to the electorate's disillusionment with neoliberalism is that their governments have deepened their reliance on primary exports (e.g. gas and lithium in Bolivia, iron ore, oil and agriculture in Brazil, and oil in Ecuador and Venezuela). Based on the surge in global demand for raw materials and agricultural products that lasted until the late 2000s, particularly from China's booming economy, left-wing governments of various political colours and convictions expanded their primary sector. In the Brazilian case, the country's impressive growth rates between 2004 and 2011 were based not only on industrial exports but also on the exploitation of natural resources. Although Brazil's economy is more diversified than some of its neighbours, agricultural and primary products still play a significant role in the country's exports. In 2014, 50% of Brazilian exports consisted of primary products, most significantly soy (13.26%), iron ore (7.45%) and crude oil (6.45%),[8] with China, the United States (US) and the European Union (EU) as its principal export markets. In this scenario, hydropower has facilitated the expansion of production and the exploitation of natural resources as well as driving the increase in domestic energy consumption.

Predominantly a state-led project under left-wing governments in Latin America, this "neo-extractivist" development model has involved using natural resource revenues to promote economic and social development.[9] This framework has involved both traditional extractivist activities, such as mining, and recent concerns such as the production of renewable energy and biofuels as well as the regional integration of infrastructure.[10] In contrast to the neoliberal drive to withdraw the state from the economy, neo-extractivism involves a 'reclaiming' and rebuilding of state authority in areas such as natural resources and industrial policy with the intention 'to oversee the construction of a new social consensus and approach to social welfare'.[11] Due to this political commitment to social development through state intervention, 'the practice of extractivism [has become] associated with an imagined national interest' in which 'the exploitation of nature serves to secure national development and sovereignty, to reduce poverty, increase social participation, to diversify local economies and to guarantee political stability'.[12] Hence this dynamic has permitted left-wing governments across the region to argue that the improvement of socio-economic conditions through government policies necessitates the expansion of natural resources exploitation.

The renewed political significance of natural resources, however, has signified that economic development priorities often override other concerns, such as democratic participation in decision-making, human rights and environmental protection. In Hogenboom's view, the 'repoliticisation' of extractivism has meant that progressive regimes 'prefer to keep [natural resource extraction] a highly centralised field of governance and to pose strict limits to local demands from civil society'.[13] For Eduardo Gudynas, the region's progressive reliance on natural resources does not resolve structural inequalities but merely compensates for the negative effects of the commodification of nature,[14] while simultaneously politically marginalising groups directly affected by this process, particularly indigenous people.[15] In Gudynas' view, the strategic importance of natural resources has also resulted in Latin American governments engaging with environmental issues 'at a surface level', leading to an environmental agenda 'that effectively incorporates actions that are functional to economic growth and a relationship to the global economy that relies on the export of primary commodities'.[16] Development strategies based on the commodification of nature have also become an object of intense contestation. According to Svampa, the Latin American 'commodities consensus' has led to an 'eco-territorial' turn in which a wide variety of socio-environmental activists have argued that natural resources, such as land, water, and forests, should not be considered as strategic resources to be commodified for development but as part of the natural, social and cultural heritage of humankind.[17]

Because of these contradictions, the politicisation of natural resources in Latin America has generated two clashing discourses: one argument – usually presented by the region's progressive governments – is that millions of ordinary citizens can benefit from the exploitation of natural resources that underpins an expansionist economic and social development agenda. The 'developmental illusion'[18] among Latin American governments reflects the view that trade in natural resources opens up opportunities to produce social and economic development, while escaping the fluctuations in global demand for primary goods. The other argument of those opposed to unfettered natural resource exploitation points to the uneven distribution of the costs and benefits and the denial of space for differing views to contest this development strategy. The affected communities have viewed the exploitation of natural resources as creating irreversible damage, which is relevant not only locally but also nationally and globally, as their views increasingly resonate with transnational environmental activists and global concerns about the state of Latin America's natural environment.

## Renewable energy as a source of conflict in Brazil

In the debate about the politicisation of natural resources, hydro-electric dams are a particularly poignant case because they not only produce two-thirds of Brazil's electricity and generate renewable energy – a cornerstone of climate change policy – but they also cause irreversible social and ecological damage according to the dams' opponents. In effect, the government argument pitches the interest of millions of Brazilians in national progress and poverty reduction against a supposedly localised discourse about the rights and specific interests of those directly affected by extractivist projects. However, local communities, together with a transnational network of activists, have offered a different conception of sustainability, one focused on maintaining indigenous peoples' distinctive spiritual relationship with their land as connected with a responsibility for the protection of nature.[19] Nevertheless, as the discourses used in conflicts about hydro-electric dams in Brazil illustrate, the politicised nature of natural resources as a highly strategic sector has simultaneously constrained opposition and opened up new domestic and international spaces for contestation.[20]

The expansion of infrastructure and hydro-electric power generation in the Amazon region[21] has formed a cornerstone of Brazil's development agenda since the 1980s. Accelerating in the 2000s in response to electricity blackouts and growing demand, the dual objective of this expansion has been to decrease the country's dependence on energy imports and to increase renewable energy sources for domestic consumption. To illustrate the importance of renewable energy, in 2014 a total of 83.7% of Brazilian electricity was generated from renewable sources, including 67.6% from hydropower. In 2014, dams located in the northern Brazilian states generated 14% of this electricity and the Ministry of Mines and Energy expects this proportion to increase to 23% by 2024 to cope with growing consumer and industrial demand. Combined, the highly controversial Belo Monte and São Luiz do Tapajós dam complexes in the Amazonian state of Pará represent 68% of the planned expansion of hydropower over the next decade.[22] The government's argument concerning dams has focused on the essential role of hydro-electricity in national development, which in turn is expected to lead to improved Brazilian living standards. For example, in response to questions about indigenous rights during Brazil's Universal Periodic Review at the Human Rights Council in 2012, the Brazilian representative pointed out that the country's 'development projects contribute not only to economic growth, but also the creation of clean energy, which accounts for a large part of the country's supply. Moreover, infrastructure creates regional and local benefits.'[23] From this perspective, hydro-electric power not only underpins national and regional development but as a renewable source of energy also contributes to Brazil's strategy to champion climate change policy.[24]

Anti-dam campaigners have challenged the government argument about the benefits of hydro-electric dams, emphasising the costs in social, cultural and environmental terms. Based on site visits to, and testimonies from, communities affected by dams in the 2000s, a 2010 Brazilian government report[25] detailed the human rights violations associated with dam construction. In addition to population displacement due to flooding, the report signalled that local communities experienced changes in traditional land use including: pollution; territorial conflicts and invasions or damage to protected areas; the loss of sacred sites; a deterioration in their quality of life; health problems; threats of violence; actual violence; and food insecurity, such as the loss of fisheries.[26] Compounding these negative effects, the Amazonian region, where most dam projects are concentrated, dominates the statistics of killings, threats and arrests of protestors associated with environmental conflicts.[27] By emphasising the tensions between these human rights violations, and the national development and environmental agendas as a strategic focus, the human rights framework has provided access to domestic and international legal instruments to challenge government policy.

## Human rights in socio-environmental conflicts

While environmental human rights are hard to define,[28] framing environmental problems in terms of human rights has allowed socio-environmental activists to appeal to internationally recognised procedural and substantive rights, particularly when struggles involve indigenous peoples. Apart from the core international human rights declarations,[29] specific principles relevant to the rights of indigenous people can be found in the International

Labour Organization's (ILO) Convention 169 on indigenous rights and the 2007 UN Declaration on the Rights of Indigenous Peoples (UNDRIP). Both ILO Convention 169, ratified by Brazil in 2002, and UNDRIP outline indigenous peoples' right to self-determination (respectively, in the Preamble and in Article 3) and cultural integrity (in Article 5 of ILO 169 and in Articles 8, 11–13, 31 of UNDRIP). Another significant element of the convention and UNDRIP is that 'indigenous peoples bear both substantive and procedural land and resource rights'.[30] From this perspective, substantive rights include the legal recognition of territories and indigenous peoples' right to use this land as well as its resources. Procedural rights include the right to participation and consultation, particularly the principle of free, prior and informed consultation when development projects affect indigenous livelihoods. ILO Convention 169 (Article 4) stipulates that governments should adopt special measures to protect indigenous groups, underlining the importance of the procedural principles of consultation and participation. However, in practice the extent to which these rights are recognised and upheld has depended on the national political context, where indigenous groups' views and interests have often been marginalised or excluded from the debate.

A human rights perspective highlights the unequal way in which people experience the impact of environmental problems, which has often translated into uneven domestic access to consultation and the legal protection of vulnerable groups.[31] These inequalities are mutually reinforcing, as the effects of environmental problems caused by natural resource exploitation tend to be more severe when local communities' livelihoods rely on land and water resources and when these communities have cultural and religious connections to the local environment, as in the case of indigenous communities.[32] These communities often lack the resources to cope with environmental damage or the capacity to pressurise their government for protection, as in the case of Brazil.

Even though the Brazilian regulatory structure provides for consultation and participation in the licensing of dam projects,[33] the consultation process has transformed into a struggle about the recognition of the rights of those affected by dams.[34] As Hochstetler argues, the environmental licensing process in Brazil is a significant source of contention because the process is open to public participation, and includes social as well as environmental considerations.[35] However, the extent to which the consultation process can effectively address substantive concerns is questionable. At a hearing about human rights violations in the Belo Monte case in 2011, a participant observed: 'Whoever went to the government hearings did not get a response to their questions. In addition, they organised hearings at the last minute so that we could not participate. We are not being heard by the companies and the government. They come, throw a book detailing the phases of the works on our table or push it under our door, and they say that this is a dialogue.'[36] In the case of the São Luiz do Tapajós Dam, in Fearnside's view the environmental licensing process 'ignores many serious socioeconomic impacts and minimises others', concluding that they are merely a figleaf to justify decisions that have already been made.[37] Similarly, Zhouri points out that Brazil's licensing procedures, while ostensibly promoting participation, have effectively depoliticised the consultation process by limiting the right to object to large development projects.[38] As the Brazilian situation shows, no matter how important procedural rights are for vulnerable groups,[39] procedures are not always effective and they do not always help to protect substantive rights or compensate for losses,[40] which has further intensified socio-environmental conflicts. While effective opposition through participation in decision-making processes has become limited, activists have turned to the international human rights framework to challenge the domestic consultation and mitigation procedures.

Reflecting the potential of the human rights framework to address these inequalities, the legal mobilisation framework posits that the primary strength of a human rights appeal not only lies in access to legal instruments but also in the power of human rights claims to be a political resource with the potential for political mobilisation and change.[41] However, the mere existence of laws and regulations has not necessarily guaranteed satisfactory outcomes, as the Brazilian case illustrates. Legal frameworks are therefore a double-edged sword as the 'law often significantly supports prevailing social relations as well as provides limited resources for challenging those relationships'.[42] As Burdon warns, while environmental human rights can provide protection, 'they have not been designed to address the underlying root causes of environmental harm'.[43] For Adelman, the 'key lies in translating human rights as aspirations or moral claims into enforceable demands',[44] which is often a difficult task. According to McCann, although the 'law provides both normative principles and strategic resources for the conduct of social struggles',[45] the strategic use of the law as a political resource needs be accompanied by political mobilisation strategies to put pressure on the political process, such as demonstrations, lobbying, transnational networking, occupations and strikes.

As discussed in the next section, the opponents of dam projects have used a range of mobilisation strategies but because they have viewed the domestic consultation procedures as lacking, they have appealed to internationally agreed human rights such as the right to information, participation and consultation, and access to justice mechanisms. As a result, international institutions[46] such as the Organization of American States (OAS)[47] and the United Nations (UN) human rights framework have increasingly become a focal point for socio-environmental campaigners in Brazil. The inter-American human rights framework not only recognises the distinctive position of indigenous peoples but has also established the relationship between environmental problems and human rights, which opens up the procedures to non-indigenous groups affected by development projects: 'where environmental degradation is not managed and minimised, it can threaten living conditions and even life itself', which means that 'human life is threatened just as human lives can be threatened by torture, imprisonment, and forced labor'.[48] The Inter-American Court of Human Rights (IACHR) can recommend precautionary measures, which has taken place in response to several Brazilian cases of social-environmental conflict, notably the Belo Monte dam but also when the state failed to protect indigenous groups whose territories had not yet been recognised.[49] Upholding procedural rights can also function to enhance democratic accountability from the OAS viewpoint, as 'governments strengthen their democratic base at the same time that they promote sustainability',[50] which has the potential to bridge the gap between the generalised benefits of extractivism and socio-environmental costs. The intersection between environmental sustainability and human rights in this definition has been interpreted in the inter-American system as the positive obligation of governments to address environmental damage and to protect indigenous territories,[51] which is reflected in the IACHR ruling on the Belo Monte dam in 2011. However, as the next section argues, the outcome of socio-environmental struggles – even when international human rights are mobilised – continues to depend on political dynamics as signified by the symbolic politics of natural resources.

### Contesting dams: human rights discourses in socio-environmental conflicts in Brazil

The campaigns against the Dardanelos and Belo Monte dams in Brazilian Amazonia illustrate how anti-dam activists have mobilised the power of human rights discourses to challenge the ethical, legal and political dimensions of the government's development agenda. In the case of the Dardanelos dam, the project faced particularly strong opposition from local indigenous communities who struggled for any recognition of the effects of the dam on their livelihoods. However, when the Dardanelos dam began operating in 2011, it soon became clear that the drive to expand hydro-electric power would override these concerns. In the Belo Monte dam case, activists therefore mobilised the symbolic and legal power of human rights by drawing international attention to Brazil's human rights record and bringing the case to international bodies such as the IACHR and HRC. As the Belo Monte dam became an international emblem of the anti-dam movement, the symbolic power of the human rights agenda pointed to the human cost of natural resource exploitation, which in turn allowed activists to frame the resultant damage in terms of nationally and internationally defined rights. At Rio+20, connecting the international human rights and environmental sustainability agendas to hydro-electric dams meant that socio-environmental campaigners could challenge the government's development agenda which, while not halting dam construction, resulted in long delays in the dam construction process.

### *Domestic contestation of the Dardanelos dam*

The Dardanelos hydro-electric dam in the municipality of Aripuanã in the north of Mato Grosso state was one of the first government-funded dam projects in the 2000s. Although the project received less attention compared to high-profile cases such as Belo Monte and Jirau, the controversies surrounding its construction are emblematic for other socio-environmental conflicts. Construction commenced in 2007 in the midst of lengthy legal and political battles that had started two years previously, exemplifying not only the problems associated with dams but also the challenges raised by opponents. Before construction began, government authorities stated that the dam would only affect indigenous territories in the region indirectly because of its location outside indigenous land.[52] Because of irregularities and omissions in the environmental impact assessment (EIA),[53] the state's public prosecutor initiated legal action against a number of companies involved in the dam's construction in 2005, demanding the cancellation of the EIA[54] and suspending the project tender.[55] The prosecutor criticised the EIA for not consulting the State Environmental Council, for not including the impact outside the municipality, and ignoring alternative locations. Neither did the assessment take into account electricity transmission to the national grid, which would cause damage over a larger area than predicted.[56] Furthermore, a 2004 report on local indigenous communities signalled the threat of migrants to the region in search of construction jobs, which would increase illegal fishing, hunting and logging, as well as polluting the natural environment.[57] The Dardanelos case therefore signals the significant role of regional public prosecutors who have challenged various aspects of the licensing and construction process, demonstrating the key role of the Brazilian legal system in the opposition to dams.

After the initial suspension of the project tender, the local public prosecutors continued to challenge the project, including the construction of electricity transmission lines, all based on irreversible social and environmental damage.[58] Despite claims that indigenous communities would not be affected, the construction process directly threatened indigenous sacred and ancestral sites, as in 2010 when the Aguas da Pedra construction company blew up an indigenous cemetery.[59] In response, around 400 indigenous activists occupied the construction site in 2010, holding construction workers hostage to demand compensation for their losses, but the damage was irreversible.[60] Similar to the other Amazonian dams,[61] opponents also questioned official information about the economic viability of the dam, particularly the effect of seasonally fluctuating water levels, raising concerns about the dam's productivity during the dry season. The Dardanelos case illustrates that opposition strategies focused on the Brazilian legal system in a struggle to challenge the narrow definition of the dam's impact but also that national legal provisions were insufficient to safeguard the rights of the groups involved.

### The Belo Monte dam on the international stage

While the Dardanelos conflict received some international attention, the protests' dynamics were primarily domestic. Instead, the Belo Monte dam has become an international symbol of the resistance of indigenous people to the damage caused by hydro-electric dams, illustrating how campaigners used symbolic and legal strategies to challenge the project. Planning for the Belo Monte dam began under the military government in the mid-1970s but accelerated in the 1990s, until Congress and the Senate finally approved the dam in July 2005. However, questions of transparency, legality and the consultation of indigenous groups continued to spark controversy,[62] eventually leading to legal challenges conducted through international human rights institutions.

Although an exhaustive overview of the legal issues surrounding the Belo Monte dam is beyond the scope of this article,[63] the complexities of the legal process explain why activists have appealed to international legal frameworks to contest the dam. At a national level protests have focused on the government's attempts to simplify the process of environmental licensing and, as in the Dardanelos case, protestors also challenged the impact studies conducted. Their argument focused on the gap between the right to consultation and the political reality, which effectively limited this right. The anti-dam coalition disputed the official EIA in 2009, pointing out omissions and irregularities. Challenging the official accounts of the impact of dams on local communities therefore turned into a key strategy to halt construction. The 230-page alternative report argued that the EIA underestimated the number of people and land area affected by the dam, overstated the project's environmental sustainability and lacked appropriate mitigation mechanisms.[64] The dam would lead to extensive flooding of towns and villages in Pará state, as well as drying up of parts of the Xingu River. At the same time, the dam offered few real benefits to local communities as the energy generated locally would need to be transported to the national grid, supporting wealthier parts of the country. While the dam displaced approximately 35,000 people during and after construction,[65] Brazilian government reports indicated that the Belo Monte dam would facilitate the construction of additional dams further upstream, with significant potential for additional displacement.[66] The alternative report also contested the often-cited argument that the actual infrastructure would not be located on indigenous land so consultation of indigenous people was not required. The official EIA had excluded rivers and shores from indigenous territories, which would significantly alter the potential effects.[67] Following these debates, in April 2010 a federal judge in Pará suspended the project tender, citing a lack of consultation of indigenous people, which is unconstitutional according to Article 231. The president of the state's Regional Federal Tribunal overturned this decision on the same day, arguing that there was no immediate danger to indigenous people because construction would not start immediately.[68]

To address the difficulties of proceeding through the domestic legal system and the lack of effective consultation, a coalition of groups from the Xingu River Basin issued a complaint with the IACHR in 2011, which became one of the most controversial international legal challenges to Brazil's development agenda. The court's ruling in April 2011 recommended precautionary measures, calling for the Brazilian government to halt construction until indigenous people were properly consulted and appropriate measures to guarantee the protection of their livelihoods were tabled. The recommended precautionary measures involved measures to protect the life and integrity of indigenous communities affected by the dam while suspending the licensing and construction process. The ruling stipulated that these measures reflected Brazil's international obligations and recommended translating the EIA into indigenous languages as well as developing a plan for the protection of peoples in voluntary isolation.[69]

Although the ruling strengthened the recognition of the rights of those affected by Belo Monte and the legitimacy of their claims in terms of Brazil's international human rights obligations, the recommendations met with a sharp response from the Brazilian government, indicating sensitivity to international pressure despite government resistance. The Ministry for Foreign Affairs responded 'with perplexity' to the demands of indigenous communities 'supposedly threatened' by the construction of Belo Monte.[70] The Brazilian government subsequently withdrew its ambassador to the OAS as well as suspending its financial contribution of US$800,000 to the IACHR. The government also responded with a 52-page document, questioning the scope of the court's powers over domestic matters and arguing that the government had taken appropriate measures to consult and protect indigenous people.[71] The government refused to recognise the jurisdiction of the IACHR in this matter, arguing that its role was subsidiary and that the consultation of indigenous people was an exclusively domestic concern as guaranteed by Article 231 of the Brazilian Constitution.[72] The document spent much time outlining indigenous policies in Brazil with little detail provided about the actual effects of Belo Monte. Tellingly, the response defined electricity as essential for fundamental development goals, such as to 'promote human dignity, guarantee national development, eradicate extreme poverty and marginalisation, and to reduce social and regional inequalities'.[73] Under pressure from Brazil, the IACHR issued a heavily toned down and revised ruling in July 2011, focusing on the protection of the health and cultural integrity of indigenous peoples as well as recommending measures to mitigate the impacts of Belo Monte.[74] Before the ruling was revised, the Brazilian Institute of the Environment and Renewable Natural Resources (IBAMA), in charge of monitoring the environmental impact of energy projects, had already issued a licence authorising construction to begin in June 2011.[75] The Brazilian response to international pressure to uphold its human rights commitments illustrates the political power of the government's counter-argument, which ended up overriding human rights concerns. However, the strong Brazilian reaction to the ruling also signals the political and symbolic significance of the international recognition of the rights of the communities involved, lending legitimacy to and recognition of campaigners' demands.

Following the IACHR ruling, in March 2012 the ILO called on the Brazilian government to observe Convention 169, requiring the consultation and participation of indigenous people regarding issues that affect their livelihoods. The tripartite commission noted that while the hydro-electric dams may not be located on indigenous lands, the former could alter 'the navigability of rivers, flora and fauna and climate, [going] further than the flooding of lands or the displacement of the peoples concerned'.[76] The legal battles also continued in Brazil. In 2012 the Brazilian Regional Federal Court halted construction of the Belo Monte dam,[77] which the Federal Supreme Court overturned again in August of the same year, after which construction resumed immediately.[78] While the construction process advanced, the Belo Monte activists decided to use the occasion of the Rio+20 sustainable development summit in June 2012 to emphasise the discrepancies between Brazil's international environmental credentials and the government's domestic development agenda. In an example of mobilising symbolic political discourse,[79] the campaigners challenged

Brazil's international reputation while the case also illustrates the limitations of a human rights strategy in a restrictive political context.

### *Symbolic politics at Rio+20*

The Rio+20 sustainable development conference in June 2012 provided the anti-dam campaigners with a platform to focus their protests on Brazil's international reputation as a champion of sustainable development. The Xingu Vivo movement organised a parallel meeting close to the Belo Monte dam in Altamira, called 'Xingu+23' to mark 20 years since the first 'Encounter of the Indigenous People of Xingu', which took place in 1989. On 16 June 2012, anti-dam activists occupied a temporary dam near Santo Antonio and opened a small channel to allow the Xingu River to flow. They also formed the words 'Pare Belo Monte' (Stop Belo Monte) on the dam, captured on an aerial photograph.[80] Another 'human banner' on the beach in Rio de Janeiro read 'Rios para a vida' (Rivers for Life).[81] Anti-dam groups featuring Brazilian and international participants held several meetings in the civil society arena in Aterro do Flamengo, located about 30 kilometres from the official conference in Riocentro. However, while Rio+20 should have underlined Brazil's global leadership ambitions as South America's largest democracy, the spatial organisation of the conference also served to illustrate the distance between civil society and the official negotiators.[82]

With the eyes of the world on Brazil, the anti-dam activists also protested against the Belo Monte dam in Rio de Janeiro. On 18 June 2012, about 1,000 indigenous protesters marched from the People's Summit in Flamengo along some of the city's busiest roads to the headquarters of Brazil's National Development Bank (BNDES).[83] In a sign of crossover between legal and political mobilisation strategies, some of the protesters held signs proclaiming that the Brazilian government should respect ILO Convention 169. They protested against the bank's role in financing large infrastructural projects, wearing 'typical clothes, bodies painted, holding *tacapes* [indigenous weapons], bows and arrows'.[84] A couple of days later, on the first day of the official Rio+20 conference, around 2,000 activists gathered outside the Riocentro conference centre. Led by the Kayapó leader, Cacique Raoni, the protesters wanted to present a document detailing indigenous demands to UN Secretary-General Ban-ki Moon.[85] They eventually talked to Gilberto Carvalho, President Dilma Rousseff's chief of staff, who allowed a small delegation to enter the official conference the next day.[86] Although the delegation met again with Carvalho at Riocentro, who conceded that there had been a lack of advance consultation in the case of Belo Monte, the delegation's presence at the official negotiations unsurprisingly did not result in the government suspending construction,[87] yet by recognising the Brazilian government's sensitivity to negative international publicity in the area of human rights and the environment, the activists very consciously engaged in symbolic politics to draw attention to their struggle.

With Brazil's environmental credentials at stake, the anti-dam protesters used the momentum of Rio+20 to highlight the tensions between human rights and development policy while Brazil was in the international spotlight, providing them with symbolic leverage in their attempt to hold the Brazilian state accountable. The Dardanelos and Belo Monte case studies illustrate the significance of combining political mobilisation with legal strategies in an attempt to force the Brazilian government to recognise the problems caused by dams. The campaigns also underline the power of the national developmental discourse, which continued to restrict the debate about procedural and substantive issues associated with large development projects. The internationally focused human rights strategy developed by the campaigners did affect the government's concerns about Brazil's international reputation, particularly when campaigners exploited the political opportunity of Rio+20 and as illustrated by the government's reaction to the IACHR ruling about Belo Monte. While Brazilian dam conflicts are ongoing, the outcomes suggest that international human rights strategies, while unable to halt dam construction or resolve substantive problems, can be employed effectively to exert political pressure in a polarised context.

## Conclusion

Weeks before the climate talks began in Paris in November 2015, Brazil experienced one of the most severe environmental disasters in the country's recent history. In the state of Minas Gerais, a mining dam being used to hold waste from iron ore extraction collapsed, releasing a torrent of toxic mud which swallowed the small town of Bento Rodrigues, killing 19 people. The disaster's human and environmental damage once again underlines the cost of natural resource exploitation while at the same time iron ore continues to be one of Brazil's most important export products. In the same month, and only days before the Paris talks started, the Brazilian environmental agency IBAMA granted the licence necessary to start filling the reservoirs of the Belo Monte dam, despite a letter from FUNAI indicating that many social and environmental conditions had not been met.[88] These examples illustrate the relationship between development, environmental and human rights priorities, · where the dominant discourse about the necessity of natural resources for national development has created a highly politicised environment. The cases discussed in this article also highlight the intensity of socio-environmental conflicts in Brazil in a context where the politicisation of energy leaves little space for effective domestic opposition, contributing to the decision to appeal to the international human rights discourse.

An analysis of socio-environmental conflicts in terms of human rights illuminates how campaigners have translated the human rights agenda into a powerful moral and political critique of development projects, supported by a wide range of other protest strategies. At the same time, the legal mobilisation approach highlights the limitations of international human rights strategies to pursue political change. Legal strategies are a double-edged sword: domestically, while the Brazilian legal system and the environmental licensing process have opened up opportunities for contestation, the national political dynamic has prevented the extension of the right to consultation to the right to object to dams. While international rulings in favour of campaigners can strengthen the legitimacy of their claims and the recognition of their rights, enforcement depends on a government's willingness to recognise these decisions. As the Brazilian case demonstrates, when an issue is as politicised and significant as energy policy, human rights norms also become politicised, thus the incorporation of these norms into national legislation and practice is by no means a linear or straightforward process.

Consequently, the political dynamic in Brazil illustrates the problematic relationship between procedural and substantive human rights in socio-environmental conflicts. Legal challenges have focused on procedural rights, particularly the right to prior consultation, which has turned into a struggle about whose costs and benefits are important when considering the impact of dams. The dam campaigners' critique has involved a struggle to recognise both the procedural and substantive rights of those involved in socio-environmental conflicts, as the official definition of the dams' impact often proved to be exceedingly narrow. The Brazilian experience also demonstrates that these procedures have often proved to be lacking, not only because of the narrow definition of the costs and benefits but also because dam construction often proceeded while the conditions set out in official impact studies had not yet been fulfilled. Furthermore, the procedural framework did not necessarily resolve substantive issues about the livelihoods of local communities and inequalities in terms of how people experience dams' effects. While appropriate procedures are essential because they offer access to information, consultation and potentially justice, to marginalised groups they do not always answer more fundamental questions about the human rights implications of irreversible damage to people's livelihoods.

## Disclosure statement

No potential conflict of interest was reported by the author.

## ORCID

*Marieke Riethof* ⓘ http://orcid.org/0000-0002-8297-5471

## Notes

1. Fundação Nacional do Índio (FUNAI), Relatório da expedição Estação Ecológica Mujica Nava/Serra dos Três Irmãos para levantamento de índios isolados, ref. no. 12, Porto Velho: FUNAI, 2009.
2. Leonardo Sakamoto, 'Governo insere 52 nomes na "lista suja" do trabalho escravo', 30 December 2001, http://blogdosakamoto.blogosfera.uol.com.br/2011/12/30/governo-insere-52-nomes-na-lista-suja-do-trabalho-escravo/ (accessed 23 March 2015).
3. Bruno Fonseca and Jessica Mota, 'Trabalhadores reféns em obras bilionários na Amazônia', Diario Liberdade, 13 December 2013; Roberto Véras, 'Brasil em obras, peões em luta, sindicatos surpreendidos', *Revista Crítica de Ciências Sociais* 103 (2014): 118.
4. The police force in question, the Força Nacional (FN), was established in 2004 as a federal force – police forces in Brazil are organised at state level – to coordinate action in case of public security emergencies. Critics have argued that the FN, being directly controlled by the president, does not have the same level of democratic control as other police forces and that its mandate now includes security at private enterprises, such as the Belo Monte dam. See Ciro Barros, 'Pela Ordem', APública, 25 April 2014, http://apublica.org/2014/04/pela-ordem/ (accessed 28 April 2016).
5. On the scope of the networks involved in socio-environmental activism in Brazil, see Kathryn Hochstetler and Margaret E. Keck, *Greening Brazil: Environmental Activism in State and Society* (Durham, NC: Duke University Press, 2007), Ch. 5; Franklin D. Rothman and Pamela E. Oliver, 'From Local to Global: The Anti-Dam Movement in Southern Brazil, 1979–1992', *Mobilization* 4, no. 1 (1999): 41–57; Philip M. Fearnside, 'Dams in the Amazon: Belo Monte and Brazil's Hydroelectric Development of the Xingu River Basin', *Environmental Management* 38, no. 1 (2006): 16–27; Eve Z. Bratman, 'Passive Revolution in the Green Economy: Activism and the Belo Monte Dam', *International Environmental Agreements* 15, no. 1 (2015): 61–77.

6. David Harvey, 'The "New" Imperialism: Accumulation by Dispossession', *Socialist Register* (2004): 75.
7. Barbara Hogenboom, 'Depoliticized and Repoliticized Minerals in Latin America', *Journal of Developing Societies* 28, no. 2 (2012): 136–7.
8. Ministério do Desenvolvimento, Secretaria de Comércio Exterior, Exportação brasileira, produto por fator agregado (Brasília: Ministério do Desenvolvimento/SECEX, 2015).
9. Alberto Acosta, 'Extractivism and Neoextractivism: Two Sides of the Same Curse', in *Beyond Development: Alternative Visions from Latin America*, ed. Miriam Lang and Dunia Mokrani (Amsterdam: Transnational Institute, 2013): 71–2.
10. Maristella Svampa, '"Consenso de commodities" y lenguajes de valorización en América Latina', *Nueva Sociedad* no. 244 (2013): 34–5.
11. Jean Grugel and Pía Riggirozzi, 'Post-Neoliberalism in Latin America: Rebuilding and Reclaiming the State after Crisis', *Development and Change* 43, no. 1 (2012): 2–3.
12. Hans-Jürgen Burchardt and Kristina Dietz, '(Neo-)extractivism – A New Challenge for Development Theory from Latin America', *Third World Quarterly* 35, no. 3 (2014): 470.
13. Hogenboom, 'Depoliticized and Repoliticized Minerals', 152; for the Brazilian case, see also Anthony Hall and Sue Branford, 'Development, Dams and Dilma: The Saga of Belo Monte', *Critical Sociology* 38, no. 6: 856–7.
14. Eduardo Gudynas, 'Estado compensador y nuevos extractivismos: Las ambivalencias del progresismo sudamericano', *Nueva Sociedad* 237 (2012): 128–46.
15. Helle Abelvik-Lawson, 'Sustainable Development for Whose Benefit? Brazil's Economic Power and Human Rights Violations in the Amazon and Mozambique', *International Journal of Human Rights* 18, no. 7–8 (2014): 801–3.
16. Eduardo Gudynas, 'Climate Change and Capitalism's Ecological Fix in Latin America', *Critical Currents* 6 (2009): 40.
17. Svampa, '"Consenso de commodities"', 41.
18. Ibid., 35.
19. Deborah McGregor, 'Living Well with the Earth: Indigenous Rights and the Environment', in *Handbook of Indigenous Peoples' Rights*, ed. Damien Short and Corinne Lennox (Abingdon and New York: Routledge, 2016), 169.
20. Kathryn Hochstetler, 'The Politics of Environmental Licensing: Energy Projects of the Past and Future in Brazil', *Studies in Comparative International Development* 46 (2011): 353.
21. For a map of the planned expansion of hydro-electric power generation, see Ministério de Minas e Energia, Secretaria de Planejamento e Desenvolvimento Energético, Plano decenal de expansão de energia 2024 (Brasília: Ministério de Minas/PPE, 2015), 392, http://www.epe.gov.br/PDEE/Relat%C3%B3rio%20Final%20do%20PDE%202024.pdf (accessed 28 April 2016).
22. Ministério de Minas, Plano decenal de expansão, 82–4, 95.
23. Human Rights Council (HRC), *Report of the Working Group on the Universal Periodic Review: Brazil, A/HRC/21/11* (Geneva: Human Rights Council, 2012), 14.
24. Marieke Riethof, 'Brazil and the International Politics of Climate Change: Leading by Example?', in *Provincialising Nature: Multidisciplinary Approaches to the Politics of the Environment in Latin America*, ed. Michela Coletta and Malayna Raftopoulos (London: ILAS, 2016).
25. Secretaria de Direitos Humanos (SDH), *Comissão Especial 'Atingidos por Barragens'* (Brasília: Secretaria de Direitos Humanos, 2010), 15.
26. See Fiocruz, 'Mapa de conflitos envolvendo injustiça ambiental e saúde no Brasil', Fundação Osvaldo Cruz, http://www.conflitoambiental.icict.fiocruz.br/ (accessed 25 August 2015); Comissao Pastoral da Terra (CPT), *Conflitos no campo Brasil 2013* (Goiânia: CPT, 2014), 77; Philip M. Fearnside, 'Brazil's São Luiz do Tapajós Dam: The Art of Cosmetic Environmental Impact Assessments', *Water Alternatives* 8, no. 3 (2015): 378–81. For the case of Belo Monte, see Secretaria de Direitos Humanos (SDH)/Conselho de Defesa dos Direitos da Pessoa Humana, Relatório de impressões sobre as violações dos direitos humanos na região conhecida como 'Terra do Meio' no Estado do Pará, Comissão Especial 'Terra do Meio', Brasília, November 2011, http://www.xinguvivo.org.br/wp-content/uploads/2012/03/Relat%C3%B3rio-CDDPH.pdf (accessed 21 November 2015), 17–18.
27. SDH, *Comissão Especial*, 16; CPT, Conflitos no Campo. On the criminalisation of protest in Brazil in recent years, see Article 19, Brazil's Own Goal: Protests, Police and the World Cup (London, 2014); UN Office of the High Commissioner for Human Rights, 'Brazilian Anti-

Terrorism Law too Broad, UN Experts Warn', 4 November 2015, http://www.ohchr.org/EN/NewsEvents/Pages/DisplayNews.aspx?NewsID=16709&LangID=E (accessed 4 November 2015).

28. Sam Adelman, 'Rethinking Human Rights: The Impact of Climate Change on the Dominant Discourse', in *Human Rights and Climate Change*, ed. Stephen Humphreys (Cambridge: Cambridge University Press, 2009), 161.

29. The Universal Declaration of Human Rights, the International Covenant on Economic, Social and Cultural Rights and the Convention on Civil and Political Rights.

30. Lillian Aponte Miranda, 'Introduction to Indigenous Peoples' Status and Rights under International Human Rights Law', in *Climate Change and Indigenous Peoples: The Search for Legal Remedies*, ed. Randall S. Abate and Elizabeth A. Kronk (Cheltenham: Edward Elgar, 2013), 56–7.

31. Stuart A. Scheingold, *The Politics of Rights: Lawyers, Public Policy and Political Change* (Ann Arbor: University of Michigan Press, 2004); Henry Shue, 'Global Environment and International Inequality', in *Climate Ethics: Essential Readings*, ed. Stephen M. Gardiner, Simon Caney, Dale Jamieson, and Henry Shue (New York: Oxford University Press, 2010), 102–5; Bradley Parks and J. Timmons Roberts, 'Climate Change, Social Theory and Justice', *Theory, Culture and Society* 27, no. 2–3 (2010): 134–66.

32. UN High Commissioner for Human Rights (UNHCHR), Promotion and Protection of Human Rights: Report of the United Nations High Commissioner for Human Rights on the Sectoral Consultation Entitled 'Human Rights and the Extractive Industry', 10–11 November 2005, E/CN.4/2006/92 (Geneva: UNHCHR, 2005), 4.

33. An overview of the Brazilian regulatory framework for the management of social-environmental issues in the energy sector since 1986 can be found in the Secretaria de Direitos Humanos report on the social and human rights effects of dams: SDH, *Comissão Especial*, 18–19. Other relevant international frameworks are the recommendations of the World Commission on Dams' final report: World Commission on Dams, *Dams and Development: A New Framework for Decision-Making* (London: Earthscan, 2000); see World Bank, *The World Bank Experience with Large Dams: A Preliminary Review of Impacts* (Washington DC: World Bank, 1996), for a review of the impacts of large dam projects. Although not ratified by Brazil, the Aarhus Convention establishes the right to access to environmental information, public participation in environmental decision-making and access to justice. For a discussion on the convention's relevance for Brazil, see Valerio de Oliveira Mazzuoli and Patryck de Araújo Ayala, 'Cooperação internacional para a preservação do meio ambiente: O direito brasileiro e a convenção de Aarhus', *Revista Direito GV* 8, no. 1 (2012): 297–328.

34. The question about whose rights are recognised reflects McCann's point that social movement struggles are often about 'the very meaning of indeterminate, contradictory legal principles'. Michael McCann, 'Law and Social Movements: Contemporary Perspectives', *Annual Review of Law and Social Science* no. 2 (2006): 25.

35. Hochstetler, 'The Politics of Environmental Licensing', 353.

36. Cited in SDH, Relatório de impressões, 15.

37. Fearnside, 'Brazil's São Luiz do Tapajós Dam', 378–81.

38. Andréa Zhouri, 'From "Participation" to "Negotiation": Suppressing Dissent in Environmental Conflict Resolution in Brazil', in *The International Handbook of Political Ecology*, ed. Raymond L. Bryant (Cheltenham: Edward Elgar, 2015), 454–5. See also Amanda M. Fulmer, 'A Strange Right: Consultation, Mining, and Indigenous Mobilization in Latin America', in *The Uses and Misuses of Human Rights: A Critical Approach to Advocacy*, ed. George Andreopoulos and Zehra Arat (Basingstoke: Palgrave Macmillan, 2014), 69.

39. Sophie Thériault, 'Environmental Justice and the Inter-American Court of Human Rights', in *Research Handbook on Human Rights and the Environment*, ed. Anna Grear and Louis J. Kotze (Cheltenham, Edward Elgar, 2015), 311.

40. Fernanda C. de Oliveira Franco and Maria L. Alencar Mayer Feitos, 'Desenvolvimento e direitos humanos: Marcas de inconstitucionalidade no processo de Belo Monte', *Revista Direito* 9, no. 1 (2013): 96; see also Human Rights Council, Report of the Special Rapporteur on the Rights of Indigenous Peoples, James Anaya: *Extractive Industries and Indigenous Peoples*, A/HRC/24/41 (Geneva: Human Rights Council, 2013), 13–14.

41. Scheingold, *The Politics of Rights*.

42. McCann, 'Law and Social Movements', 25.

43. Peter D. Burdon, 'Environmental Human Rights: A Constructive Critique', in *Research Handbook on Human Rights and the Environment*, ed. Anna Grear and Louis J. Kotze (Cheltenham, Edward Elgar, 2015), 62, 73.

44. Adelman, 'Rethinking Human Rights', 167.

45. McCann, 'Law and Social Movements', 22.

46. The following are examples of UN documents focusing on human rights and extractivism: UNHCHR, Promotion and Protection; HRC, Report of the Special Rapporteur.

47. To date the Inter-American Commission on Human Rights has a Proposed American Declaration on the Rights of Indigenous Peoples (1997), while also drawing on the American Convention on Human Rights, its own resolutions regarding indigenous issues since the early 1970s, and other international human rights instruments such as UNDRIPS and ILO Convention 169. For a comprehensive overview relevant to the right to land and natural resources, see Inter-American Court of Human Rights (IACHR), *Indigenous and Tribal Peoples' Rights over their Ancestral Lands and Natural Resources*, OEA/Ser.L/V/II.Doc. 56/09 (Washington, DC: IACHR, 2009).

48. Organization of American States, Report of the Unit for Sustainable Development and Environment on its Efforts in the Field of Human Rights and the Environment, CP/CAJP-2100/03, 14 November 2003, OAS Committee on Juridical and Political Affairs, http://www.oas.org/en/sedi/dsd/ELPG/Docs/report_human_rights.pdf (accessed 5 November 2015), 2.

49. See IACHR, *Comunidades indígenas da Bacia do Rio Xingu, Pará, Brasil, Precautionary Measures 382/10* (Washington, DC: IACHR, 2011); IACHR, 'Indigenous Peoples Ingaricó, Macuxi, Wapichana, Patamona and Taurepang in Raposa Serra do Sol, Roraima State Brazil' (Washington, DC: IACHR, 2004).

50. OAS, Report of the Unit for Sustainable Development and Environment, 2.

51. UN Office of the High Commissioner for Human Rights (OHCHR) and United Nations Environmental Programme (UNEP), *Human Rights and the Environment* (Geneva: OHCHR and UNEP, 2012), 17–18; Fergus MacKay, 'The Rights of Indigenous Peoples in International Law', in *Human Rights and the Environment: Conflicts and Norms in a Globalizing World*, ed. Lyuba Zarsky (London: Earthscan, 2002), 17–18; Linda H. Leib, *Human Rights and the Environment: Philosophical, Theoretical and Legal Perspectives* (Leiden: Martinus Nijhoff, 2011), 74–5.

52. Telma Monteiro, 'Dardanelos e Belo Monte: A história se repete', EcoDebate: Cidadania & Meio Ambiente, 3 August 2010, http://www.ecodebate.com.br/2010/08/03/dardanelos-e-belo-monte-a-historia-se-repete-artigo-de-telma-monteiro/ (accessed 3 November 2015).

53. Fearnside, 'Brazil's São Luiz do Tapajós Dam', 391.

54. Hochstetler, 'The Politics of Environmental Licensing', 355.

55. Ministério Público do Estado de Mato Grosso, 'Suspensos efeitos do EIA/RIMA para construção da hidrelétrica Dardanelos', 23 September 2005, http://www.mpmt.mp.br/imprime.php?cid=40476&sid=58 (accessed 3 November 2015).

56. Bruno Fonseca and Jessica Mota, 'As pegadas do BNDES na Amazônia', APública, 15 October 2013, http://apublica.org/2013/10/investimentos-b,ndes-em-infraestrutura-na-amazonia-caso-da-hidreletrica-de-dardanelos/ (accessed 3 November 2015).

57. Gilberto Azanha, 'Estudo socioeconômico sobre as terras e povos indígenas situadas na área de influência do AHE Dardanelos, no Rio Aripuanã', Centro de Trabalho Indigenista, November 2004, http://bd.trabalhoindigenista.org.br/sites/default/files/Azanha_AHE-Dardanelos-Componente-Indigena.pdf (accessed 3 November 2015).

58. Fiocruz, 'Povos indígenas de Aripuanã lutam contra instalação de hidrelétrica', http://www.conflitoambiental.icict.fiocruz.br/index.php?pag=ficha&cod=345 (accessed 3 November 2015).

59. Zachary Hurwitz, 'Another Indigenous Tragedy Highlights the Inviability of Amazonian Dams', International Rivers Network, 29 July 2010, https://www.internationalrivers.org/blogs/258/another-indigenous-tragedy-highlights-the-inviability-of-amazonian-dams (accessed 3 November 2015); Fonseca and Mota, 'As pegadas do BNDES'.

60. Fiocruz, 'Povos indígenas de Aripuanã'; Conselho Indigenista Missionário (CIMI), 'Ocupação da usina Dardanelos: Grito aos surdos', CIMI Regional Mato Gross, 28 July 2010, http://www.cimi.org.br/site/pt-br/?system=news&action=read&id=4830 (accessed 3 November 2015).

61. In particular, concerns have focused on the dam's operational capacity during the dry season, which is expected to be 10%. For a useful summary of the debate on the economic viability of the Belo Monte dam, see Eve Z. Bratman, 'Contradictions of Green Development: Human

Rights and Environmental Norms in Light of Belo Monte Dam Activism', *Journal of Latin American Studies* 46, no. 2 (2014): 269–71.

62. Instituto Socioambiental, 'Especial Belo Monte: A polêmica da usina Belo Monte', http://www.socioambiental.org/esp/bm/hist.asp (accessed 21 February 2012); Fearnside, 'Dams in the Amazon', 20–3; Sabrina McCormick, 'The Brazilian Anti-Dam Movement: Knowledge Contestation and Communicative Action', *Organization Environment* 19, no. 3 (2006): 321–46.

63. A comprehensive analysis can be found in Hochstetler, 'The Politics of Environmental Licensing'.

64. Painel de Especialistas, Análise crítica do Estudo de Impacto Ambiental do Aproveitamento Hidrelétrico de Belo Monte, Belém, 29 October 2009, http://www.socioambiental.org/banco_i-magens/pdfs/Belo_Monte_Painel_especialistas_EIA.pdf (accessed 5 November 2015).

65. Hall and Branford, 'Development, Dams and Dilma', 854.

66. Fearnside, 'Dams in the Amazon', 3.

67. Minority Rights Group International, *State of the World's Minorities and Indigenous Peoples 2012: Focus on Land Rights and Natural Resources* (London: Minority Rights Group International, 2012), 94–6.

68. JusBrasil, 'TRF autoriza o leilão da usina Belo Monte', http://expresso-noticia.jusbrasil.com.br/noticias/2156721/trf-autoriza-o-leilao-da-usina-belo-monte (accessed 5 November 2015).

69. IACHR, *Comunidades indígenas*.

70. Itamaraty, Solicitação da Comissão Interamericana de Direitos Humanos (CIDH) da OEA, Press Release, 5 April 2011 (Brasília: Itamaraty).

71. República Federativa do Brasil, Comunidades tradicionais da Bacia do Rio Xingu, Pará: Informações do estado brasileiro, 25 April 2011, http://www.xinguvivo.org.br/wp-content/uploads/2010/10/Resposta_do_Estado_MC_030520111.pdf (accessed 6 November 2015).

72. Itamaraty, Solicitação da Comissão.

73. República Federativa do Brasil, 'Comunidades tradicionais', 24.

74. IACHR, *Comunidades indígenas*.

75. IBAMA, 'Ibama autoriza a instalação da Usina de Belo Monte', 1 June 2011, http://www.ibama.gov.br/publicadas/ibama-autoriza-a-instalacao-da-usina-de-belo-monte (accessed 21 November 2015).

76. ILO, 'Follow-up to the Recommendations of the Tripartite Committee, Indigenous and Tribal Peoples Convention, 1989 (No. 169)', 2012, http://www.ilo.org/dyn/normlex/en/f?p=NORMLEXPUB:13100:0::NO::P13100_COMMENT_ID:2700476 (accessed 3 November 2015).

77. Christiane Peres, 'Obras de Belo Monte são paralisadas até que consultas aos povos indígenas sejam feitas pelo Congresso', 15 August 2012, Instituto Socioambiental, http://site-antigo.socioambiental.org/nsa/detalhe?id=3644 (accessed 6 November 2015).

78. Christiane Peres, 'Ayres Britto acata pedido da AGU e obras de Belo Monte são retomadas', 28 August 2012, Instituto Socioambiental, http://site-antigo.socioambiental.org/nsa/detalhe?id=3656 (accessed 6 November 2015).

79. Alison Brysk, '"Hearts and Minds": Bringing Symbolic Politics Back in', *Polity* 27, no. 4 (1995): 561.

80. 'Xingu+23, 15.06: Liberando o Xingu', 17 June 2012, http://amazonia.org.br/2012/06/xingu-23-15-06-liberando-o-xingu/ (accessed 10 November 2015); Abelvik-Lawson, 'Sustainable Development', 807.

81. Author's research notes, Rio de Janeiro, 19 June 2012.

82. Ibid., June 2012.

83. For an analysis of the role of the BNDES in financing development projects, see Fonseca and Mota, 'As pegadas do BNDES'.

84. 'Protesto reúne mais de mil indígenas diante do BNDES', 19 June 2012, O Estado de São Paulo.

85. Articulação dos Povos Indígenas do Brasil (ABIP), 'Carta do Rio de Janeiro: Declaração final do IX Acampamento Terra Livre – Bom viver/Vida plena', Rio de Janeiro, 15–22 June 2012, http://blogapib.blogspot.nl/2012/06/documento-final-do-ix-acampamento-terra.html (accessed 6 November 2012).

86. The following video depicts the protests at Rio+20 and the exchange with government representatives outside Riocentro: Relatório Áudio Visual: Acampamento Terra Livre, Rio+20, June 2012, https://www.youtube.com/watch?v=c4sTminBWFY (accessed 6 November 2015); Renato Santana, 'Depois de protesto, indígenas conseguem entrar na Rio+20 para entregar

reivindicações', Conselho Indigenista Missionário, 21 June 2012, http://www.cimi.org.br/site/pt-br/?system=news&action=read&id=6348 (accessed 6 November 2015).

87. Renato Santana, 'Informe no. 1.020: Ministro Gilberto Carvalho admite ausência de consulta, mas que obra da UHE Belo Monte seguirá', Conselho Indigenista Missionário, 22 June 2012, http://www.cimi.org.br/site/pt-br/?system=news&action=read&id=6350 (accessed 6 November 2015).

88. 'Outcry as Brazil Authorizes Operation of the Belo Monte Dam', 26 November 2015, Earth First Newswire, http://earthfirstjournal.org/newswire/2015/11/26/outcry-as-brazil-authorizes-operation-of-the-belo-monte-dam/ (accessed 28 November 2015); Letter from João Gonçalves da Costa, President of FUNAI to IBAMA, Brasília, 12 November 2015, available from http://www.funai.gov.br/arquivos/conteudo/ascom/2015/img/11-nov/oficio587.pdf (accessed 28 November 2015).

# Extracting justice? Colombia's commitment to mining and energy as a foundation for peace

John-Andrew McNeish

In this article critical consideration is given to the idea that natural resource extraction can pay for peace and justice. The Colombia government claims that the mining and energy sector represents a driver for economic growth and a key source of the financing needed in a post-conflict period following the forthcoming signing of peace accords. Recent research suggests this argument is severely flawed. Legal and socio-economic dynamics indicate that the encouragement of the extractive sector does not represent a route to justice, but rather an increase in insecurities and a new phase of human rights violations in the country.

## Introduction

The Colombian government of Juan Manuel Santos claims that the mining and energy sector represents a key driver for development and economic growth.[1] Indeed, the Santos administration has in its second term and in the run-up to a forecasted peace accord with the main guerilla force in the country, the Revolutionary Armed Forces of Colombia (FARC), renewed emphasis on the extractive sector. Outside of foreign assistance, the extractive economy is seen as a vital source of the needed state income to cover the costs of an eventual peace and continued economic development.

In this article I draw on recent qualitative field research, media reports and other secondary sources to question whether this is an appropriate part-solution to over half a century of armed conflict. The research, still continuing, has been carried out during several periods since 2013 and principally in connection with the Norwegian Research Council funded *Extracting Justice* project.[2] In contrast to earlier research on the links between extraction and civil war,[3] the article stresses that the insecurities caused by the extractive economy will continue into a post-conflict period. I highlight that rather than forming the basis for equitable development in the country, the recent expansion of the extractive sector has – as is common with other petro/extractive-reliant states[4] – resulted in more economic uncertainty for both the national economy and most of the population. More significant still, as the expansion of the extraction of hydrocarbon over the last decade indicates, it has fuelled the mutation of the armed conflict and resulted in a series of humanitarian crises throughout the country including the single largest internally displaced population in the world.[5] In making this argument I do not mean to claim an automatic relationship, or resource curse, between extractive economies and *weak* or *fragile* states,[6] but rather suggest, in line with more critical thinking,[7] that the dangerous outcomes of an extractive economy are a response to legal and institutional structuring reflecting clear political motivations and decisions to empower capital at the expense of civil society.

The article furthermore argues that rather than support the conditions for the enhancement of human rights in the country, recent changes to policy aimed at easing and speeding up codes and regulations governing extractive practice have further encouraged increasing tendencies towards a breach of rights to mitigation and consultation. Whereas the current and previous Colombian presidents, Santos and Uribe, differ greatly in their approach to ending the armed conflict, I emphasise in this article that the current government has incubated, rather than opposed, the legal and political conditions needed to continue the expansion of the extractive economy. Indeed, as I emphasise in this article, encouraged directly and indirectly by further liberalisation, Colombia has witnessed in recent years the re-articulation of armed conflict in the country, facilitating the legal and illegal control of extractive installations. In response to these changes, Colombian civil society has made a series of efforts to confront extractive projects and to express claims to popular sovereignty and environmental justice through joint strategies of protest and democratic popular referenda. As I demonstrate here, these actions have been met by a combination of legal state repression in the form of legal challenges, heightened levels of securitisation and the limitation of legal rights in extractive zones. More worrying still, they continue to result in activists becoming the target of extra-judicial repression such as physical threats, violence and assassinations.

## Environmental peace?

A peace accord between the Colombian government and the FARC now looks inevitable. Talks taking place between the main warring parties in Havana since 2012 have now matured to a stage where both sides have ceased attacks[8] and agreed on a date – 23 March 2016 (negotiations continued past this date) – for the delivery of a final document to the Colombian people. Following 55 years of armed conflict, it now seems that a final peace deal will soon be presented to the people of Colombia in the form of a national referendum. At the time of writing CERAC, a Colombian think-tank that monitors conflict events, has reported that the past two months have been the most peaceful that Colombia has enjoyed since 1975.[9]

Such movement towards a final settlement has justifiably spurred a growing sense of optimism amongst the majority of the Colombian population. In the course of my research in Colombia in 2014 and 2015 it was made clear in private conversations and interviews that this optimism is not, however, shared by everyone in Colombia. Indeed, concerns about the government's vision of a post-conflict period are not only expressed by the right-wing political opposition led by Alvaro Uribe who oppose the talks and believe in the eventual success of a military solution,[10] or by the remaining left-wing guerilla National Liberation Army (ELN). Whilst supporting the need for a peace accord, public intellectuals, domestic civil society and international organisations throughout the country question whether the contents of the current agreement are sufficient.[11]

In a recent column in the national daily newspaper, César Rodríguez Garavito, a well-known human rights lawyer and member of Dejusticia,[12] posed the question whether a territorial peace as proposed by the current peace talks will be possible without delivery of an environmental peace.[13] Referring directly in his article to recent efforts by communities in the Colombian district of Tolima to confront state-backed efforts to establish a gold mine in the heart of a rich agricultural region (and to which I will return below), Rodríguez Gavarito suggested that:

> The post-conflict [following the eventual signing of a peace accord between the FARC and the Colombian government] will depend on the resolution of these kinds of socio-environmental conflicts, because the map of territory that the Government estimates as a priority for the consolidation of the peace coincides with the municipalities that are titled for mining, zones of environmental protection and problems of water. It is in these same areas that socio-environmental conflicts will increase when the peace makes them more accessible ... It therefore follows that there will be no peace without environmental peace. The government understands the first part of this equation ... but at the same time is doing everything possible to remove the power of municipalities and complicate the path of communities who are searching to construct environmental peace through legal mechanisms such as popular consultations.[14]

The suggestion that Colombia's peace process faces environmental challenges as well as political and territorial ones is a message shared by the United Nations (UN) office in Colombia. In a recent report, Fabrizio Hoschield, the Resident Humanitarian Coordinator of the UN in Colombia, correctly notes that the armed conflict has both caused environmental destruction and been a factor in the conservation of the same.[15]

The insecurities of the last decades have left large areas of the country untouched by the wider impacts of development. However, he also stresses that grave damage to the environment has been caused by the conflict. The illegal economy of the war and efforts to stop it have resulted in the spread of coca and palm-oil plantations and spraying of poisonous glyphosate over larger areas in eradication efforts. Damage has also been caused to national parks and areas of natural beauty by the spread of illegal mining and unregulated resource extraction. These spaces have often also become areas of dispute and destruction as the warring groups and criminal bands have attempted to control or tax them. Over ten years of heightened counter-insurgency and para-military activities have significantly weakened the guerrillas' strength, and have pushed their rural hideaways into ever more remote areas.[16] The remaining geographical areas of armed conflict largely coincide with areas of fertile lands of interest to agri-business and underground natural resources that are especially attractive to mining, quarrying and oil companies.[17]

Recent reports by the *Consultoría para los Derechos Humanos y el Desplazamiento* (CODHES) also make it clear that there are strong links between the expansion of illegal and legal mining and the displacement of many communities in the country.[18] Official figures recognise there to be currently over five million internally displaced people in the country.[19] A CODHES report published in 2012 found there to be a large military and paramilitary presence in mining zones.[20] Whilst the armed forces protect private investment, the paramilitaries suppress social protest and create displacement. An Indepaz (Institute of Studies for Development and Peace/Instituto de Estudios para el Desarrollo y la Paz) report published in 2012[21] importantly also reveals that 'mining has not displaced narco-trafficking, instead it serves for money laundering assets and in some cases to help with the finances of these groups'. The report highlights that the FARC and paramilitary groups such as *Los Urabeños* and *Los Rastrojos* are the ones that 'now control the illegal mining on the Pacific Coast'. In areas such as Timbiquí in Cauca and Istmia in Chocó, mining sustains armed groups operating at the margins of the law, protecting the jungle corridors through which drugs are moved.[22] Mining operations in these areas also justify the presence of armed actors as a means of providing security to prevent robberies and extortion.[23]

## Engine for change and regulation

Given the gradual mutation of the armed conflict – now as much about oil and minerals as it is coca, land and political ideology – it is clear that there is an empirical basis to doubts whether the set-up of the current peace talks will deliver a broad enough solution to ongoing violence. As some analysts suggest, this wider context indicates that there is every sign that a 'post-conflict' Colombia will be far from a 'post-violence' scenario.[24] The Havana peace talks have delivered a series of important agreements,[25] but as critics note, questions regarding the national economic model and its extractive base have consciously been avoided. Santos made it clear to the media before the peace talks started that the economic and political model of the country would not be under discussion. He suggested that if the FARC wanted this, they would have to win the right to such discussion through participation as a political party in democratic elections.[26]

Since arrival in government in 2010 the Santos government has persevered with an economic plan aimed at spurring growth through enabling conditions for five central economic driving forces, amongst them the *energy-mining locomotive (locomotora energetica-minera).*[27] Now in its second term, the Santos government continues to insist on the developmental imperative to increase large-scale and open-pit mining. Under Santos the promotion of extractive capital has accelerated significantly, mirroring the high prices for commodities in international markets until the fall in 2014/2015. In 2010, 5.8 million hectares of land were licensed for mining exploitation. Colombia is now the world's fourth largest exporter of coal and Latin America's fourth largest oil producer. In 2011 Colombia's gross domestic product (GDP) grew by 6%, largely due to the commodity boom produced by these moves. Wider hydrocarbon production has also increased. Gas production rose from 200 billion cubic feet in 1999 to 400 in 2010. Oil production has risen from 595,000 barrels a day in 2008 to 923,000 in 2011 to estimates of over a million in 2012. A modern gold rush has also taken place with production rising from 40 tons annually to 80 tons in 2012. Responding to the Santos government's statements regarding the formal inclusion and protection of the energy sector in its economic model, a series of new trade agreements have also been recently signed with the United States (US), the European Union (EU) and Israel. New investment has not only directly fed the expansion and further exploration in the oil sector, but also stimulated infrastructure construction, including the building of new roads, bridges and ports at key points throughout the country to assist the export of raw materials.

During his second presidential campaign, Santos emphasised that education, equity and peace would be the guiding principles of the new government. However, in taking office anew he has renewed his support for the importance of the extractive sector, and in particular of large-scale mining, to the national economy. A recent trip made by Santos to Europe in 2015 made clear his government's hopes that a significant proportion of the estimated US $45 billion needed for a successfully implemented demobilisation and peace process will come from foreign donors. However, official statements also make it clear that the energy and mining driver remains a key foundation for economic development in the immediate future.[28] Energy and mining account for 25% of total government revenues (up from 12% five years ago), 70% of total exports, and 55% of foreign direct investment (FDI).

Existing estimates of the successful implementation of a demobilisation process after the final signing of the accords and public referendum on its acceptance place the bill at circa US$45 billion.[29] This is around 15% of Colombia's national GDP, and the totality of the country's central bank reserves. The cost of peace is therefore a very real question, and given the prominence of mining in the current national economy it is recognised by the government as one of the key economic areas from which the state can generate needed capital. It is also an area of the economy from which the government has forecasted a rapid 'peace dividend'. Government officials stress that the signing of a peace treaty would end the losses seen in the sector as a result of the armed conflict in recent years, and encourage higher levels of investment. In 2014, over US$500 million was lost in oil revenues due to attacks on energy infrastructure by guerilla groups. The Colombian Oil Association estimated that some 4.1 million oil barrels were spilled over the last three decades.[30] At present, circa 80% of Colombian gold comes from illegal mining. Over the last five years, 2781 illegal mines have been detected, representing over US$2.5 billion of the estimated revenues of the sector. It is claimed that this would disappear following the accords, as peace would allow for the further regulation and monitoring of the sector.[31]

## The costs of the energy-mining 'express'

Despite the hopes of the Santos government that the mining and energy sector represents a guaranteed source of state revenue, there has been official recognition that the recent fall in international commodity prices will make the goal of increasing reserves difficult.[32] Responding to this changing picture, the president of the National Hydrocarbons Agency (ANH) suggested that the best way to address these instabilities is to further increase exploration.[33] However, to achieve this he has suggested in accordance with the presidency and Ministry of Mines that legislative changes are required to get rid of log-jams and to speed up environmental licensing.

In January 2015 a new decree (2041) was passed by the central government to speed up the licensing process. Commonly referred to by critics as the 'decreto de licencias express' (the express licensing decree) the government claimed that the reform would responsibly assist with reducing the time required to process new exploration licences. Although unable, following significant public opposition, to reduce, as proposed, the schedule from 90 days for new projects and 60 days for the modification of existing projects, with the decree, the government did manage to significantly simplify the internal governmental application and processing procedure. Decree 2041 establishes the precedent that instead of several submissions, all information regarding a project should be submitted on one occasion. If there are details missing the application will be immediately rejected. It further introduces the requirement for the oral presentation of the proposed project to the environmental authorities. It is thought that this will aid communication and quickly dispel misunderstandings.

The government intended these changes to reduce the bureaucracy and speed up the efficiency of processing new applications for extractive licences. However, environmental, civil society and indigenous organisations throughout the country immediately recognised that the decree would significantly reduce the bureaucratic, legal and institutional hurdles that had acted – albeit poorly – as safeguards under the previous licensing decree (2820).[34] Communities or non-governmental organisations (NGOs) opposing the granting of specific licences or questioning formal consultation processes are formally left to press their claims and complaints through the court system, including use of the constitutional court and Auditor General's office. The result of this is that whereas legal process can occur, it rarely blocks the initial steps towards licensing. This in turn places such claims and complaints at a significant temporal and legal disadvantage.

According to the 1991 Colombian constitution it is the right of every municipality in Colombia to autonomously decide on how to best preserve the environment in its territories and to regulate the use of land appropriately.[35] It further states that Colombia is a democratic, participatory and pluralistic state (article 1), which is based on the principle of participation of all (article 2) and in the principle of popular sovereignty (article 3). The emphasis on the right to participate in the local governance of the government is particularly strong in relation to the country's indigenous population.[36] Article 7 of Colombia's constitution furthermore establishes respect for an ethnically and culturally diverse population.

Aiming to establish legal protection, the constitution recognises indigenous territories as reservation lands (*resguardos*), which function as political and administrative entities. In 2001, the Colombian government changed two articles in the constitution. To guarantee the stability of state funds for social investment in territorial entities including indigenous *resguardos*,. Law 715 of 2001 was subsequently enacted, to regulate the distribution of these funds, and their use. The same law establishes that indigenous *resguardos* will receive a percentage of national funding each year, to be used for education, health, housing, drinking water and productive projects. The constitutional framework also secures indigenous land tenure and communities' right to choose their own authorities and administer their internal affairs according to customary laws. The constitution also places specific responsibility on indigenous authorities to ensure 'the preservation of natural resources'. The 1991 political constitution states regarding natural resource extraction on indigenous territory that:

Exploitation of natural resources in the indigenous (Indian) territories will be done without impairing the cultural, social, and economic integrity of the indigenous communities. In the decisions adopted with respect to the said exploitation, the government will encourage the participation of the representatives of the respective communities (article 63).

The constitution also contains the following provision regarding rights to consultation and land expropriation:

The state will protect and promote associational and collective forms of property. Due to public necessity or social interest as defined by the legislator, expropriation will be possible pursuant to a judicial determination and prior indemnification. The latter will be determined in consultation with the interests of the community and of the affected party. In cases determined by the legislator, such expropriation may occur by administrative mean, subject to a subsequent administrative legal challenge, including with respect to price.

Matching the Colombian state's ratification of ILO 169 and the UN Declaration on the Rights of Indigenous Peoples (2007), provision is then made by the constitutional court for consultation to take place before the granting of an environmental licence for extractive purposes. According to the constitutional court, providing information or notification to an indigenous community about a natural resource exploration or exploitation project does not constitute prior consultation. Finally, according to the constitutional court, prior consultation consists both of communication and understanding, characterised by mutual respect and good faith between communities and the government authorities.[37]

Even before the introduction of decree 2041, Colombian indigenous organisations had highlighted the gap between *de jure* rights and their flawed *de facto* respect and practice by the government. In 2015 a total of 156 consultation exercises and 640 agreements had been made between the government and indigenous communities in respect of the pre-existing legislation. Between 1993 and 2006, the Constitutional Court also found 18 cases where indigenous land rights were violated by 'intrusive initiatives' or large infrastructural projects. These included several cases with high international profiles due to the extensive campaigning by national and international NGOs, for example the cases of the Motilón Bari, U'wa and Embera Katio peoples. Despite these legal rulings a gap between law and its practice has persisted. According to a report (2009) by the UN Special Rapporteur on Indigenous Affairs:

> ... whilst the government provided information about consultations regarding various projects, the information does not establish that the consultations were carried out in accordance with the relevant international standards.[38]

Despite appearing to have some of the most progressive legal institutions, laws and regulations in the region, the number of legal claims of breaches of human and indigenous rights in the country have increased rather than reduced over the last decade. This is the result of the rather piecemeal manner in which changes and improvements to legal standards have been introduced over time. The result is a 'scattered number of norms, guidelines, decrees, and presidential directives, which for the time being, must serve as a compass on how to fulfill the state's duty to consult ethnic minorities. However, they provide no legal security for the 'stakeholders'".[39]

The first attempts to regulate prior consultation were made in 1993 with the enactment of Law 99, which created Colombia's first environmental authority. Along with the Ministry of the Environment and the National Environmental System, the law also established the responsibility to consult ethnic minorities as a prerequisite for accessing environmental licences in connection with extractive projects in their recognised territories. In accordance with the Colombian legal system, regulatory decrees are often enacted to implement new laws. Decree 1320 was established in 1998 to provide guidelines for analysing the impacts of natural resource extraction on indigenous and afro-descendent communities, and to establish measures that would defend their territorial integrity. Unfortunately, Colombian lawmakers failed at this time to consider an important detail when introducing new legislation for prior consultation, that is, they failed to consult the communities themselves regarding the content of the new legislation. As a result, the Constitutional Court, the guardians of fundamental rights in the country, ruled decree 1320 unconstitutional. The Constitutional Court then attempted to fill the void itself through a series of individual rulings in cases of environmental licensing. In 1997 it had, in Ruling SU039, ruled that the right to an ethnic group's participation through prior consultation was a fundamental right, and thus essential to preserving the ethnic, social, economic and cultural integrity of ethnic communities. This was reinforced much more strongly in 2008 through another ruling (CO30/2008), which clarified for the first time how administrative and legislative acts likely to affect ethnic communities were to be conducted. As a result, the Forestry Law of 2006, the Rural Development Statute of 2007 and reform of the Mining Code in 2010 were all ruled unconstitutional because of their failure to consult ethnic minorities in their creation. The court's decision T129/2011 delineated parameters with which to make it possible for ethnic communities to exercise their rights in line with the established principles for prior consultation. It also mandated securing communities' free, prior and informed consent before community members are resettled or displaced, when there is danger of social and economic impact, or when development activities pose a risk of contamination or involve the storage of toxic waste on ancestral lands. Finally it ensures the involvement of the National Ombudsman (*Defensoría del Pueblo*), the Auditor General

(*Controloría*) and the Prosecutor General's office (*Procuraduría General*) during the prior consultation process. A further tightening of the system appeared to take place under the Santos government with the introduction of 2613/2013, which lays out five specific steps for carrying out prior consultation. However, because of the rather piecemeal or organic fashion in which laws and regulations have developed in Colombia, much of this system can easily be knocked aside by the introduction of another decree, such as decree 2041.

### Legal moves and countermoves

Indigenous and afro-descendent communities throughout Colombia have made numerous statements regarding the inefficiency and failure of the system in the country to protect their legal rights, and of the continuing importance of prior consultation, for example:

> We see mining as related to human rights, but the State tells us that it is not, that it has nothing to do with human rights. We think it is. A right to territory is a human right, because it is the space where we can be ourselves. If we leave our territories we can no longer be ourselves (Carlos Eduardo Gómez Restrepo, Governor, Cañamomo Lomaprieta Indigenous Reserve, 16 May 2015).[40]

Indeed, activists have not only commented on apparent legal inconsistency, but the bureaucratic incompetence of the state to carry out what has been legislated.

> The truth is that the mining, environmental and ethnic politics of the government continues to be in chaos, and stands in contrast to the professionalism of other areas, such as health. As we have seen those of us observing consultations, the delays are a result of the lack of order and efficiency. They constantly rotate the functionaries responsible within the Ministry of the Interior; they in turn fail to coordinate with the Ministry of Mines, the Ministry of the Environment; rickety and without direction, it all leads to disharmony (Cesar Rodriguez Gavarito, El Espectador, 24 November 2015).[41]

In the face of the insecurities of the current legal and political system for prior consultation, a diversity of mechanisms have been used by the indigenous and afro-descendent communities to pressure for further change and legislative clarity. Here direct militancy parallels other attempts to use the national and international legal and political system to pressure Colombian authorities and institutions. Direct action, blockades and protests continue to take place regularly throughout the country and, in particular, in zones of natural resource extraction. In many cases indigenous and afro-descendent organisations have formed coalitions with other social actors similarly concerned and proactive in their opposition to the conditions under which resource extraction is taking place, as well the more general extractive terms of national economic policy.

Social movements in Colombia have actively employed the concept of *minga* – which translates as a form of popular sovereignty that emanates from the democratic base – as a call to action in recent years.[42] From 11 October to 24 November 2008, Colombia's popular movement, spearheaded by the country's indigenous organisations, carried out an unprecedented six-week mobilisation to protest President Alvaro Uribe Vélez's economic development and military/security policies, as well as the ongoing violations of the rights of indigenous people. The *minga popular*, as it was called, brought together upwards of 40,000 people. The *minga popular* was described by its leadership as the beginning of a nationwide 'conversation with the people', a popular uprising of sorts, designed to transform Colombian society and politics through coordinated, non-violent mobilisation. It received considerable support from the Colombian population, as well as tremendous expressions of international solidarity. There is also a recognised direct connection between the 2008 *minga popular* and the *paro agrio nacional* (national agrarian strike).

In August 2012, Colombia experienced what social activists called a *social earthquake*.[43] Thousands of indigenous peoples, small farmers and small-scale miners from municipalities throughout the country paralysed Colombia for a week through a series of strikes, protests and blockades. The protesters used these actions to highlight public rejection of the government's neoliberal economic model and in particular the terms of recent free trade agreements and the extremely liberal concession terms granted agri-business and extractive corporations. Student organisations and industrial workers gave further support to this movement of people as the protests and strikes reached the nation's cities and capital. On each of these occasions of open protest, the government claims that they respected the right of the protesters to protest, but have also sent in the heavily equipped riot police (ESMAD) to physically manage the protests and arrest their ringleaders. The heavy handed actions of the police have resulted in the injury and deaths of protesters in recent years.

Other than open militancy, indigenous, afro-descendent and environmental social movements in the country have also looked to other democratic legal and political mechanisms to both raise public awareness and to block extractive projects. Indigenous and afro-descendent organisations have been strong proponents of formalising alternative development plans for their territories. When conversing with a local leader from the *Caño-momo Lomaprieta* indigenous reserve I was told that in their negotiations with the state they insist: 'we are the state' (*nosotros somos el estado*). In the reserve, the community have developed their own territorial and development plan, and it is this that should orient the national government in consultation with the local community.

In other cases, indigenous and afro-descendent communities have joined broad cross-class social movements in campaigns to confront and stop extractive activities. Of marked importance in terms of national legal and political debates, several attempts have been made by local municipalities in response to the calls of social movements to organise popular referenda. One case that has been given significant national and international attention is described in more detail here. In 2007 the international mining conglomerate Anglo Gold Ashanti (AGA)[44] announced to the international press the discovery of a major new gold deposit in the foothills of the *Los Nevadas* mountains. In their press announcement the AGA stated that the discovery was 'one of the projected gold resources with the greatest potential in Colombia, and if possible to mine would be one of the greatest discoveries in Latin America in the last decade'. In the company's Annual Review for 2010 the project was highlighted as 'the gold industry's most significant virgin discovery of recent time'.[45] The company expects the mine to produce two to three million ounces of gold per year by 2025 by AGA.[46]

In the years that followed, local communities within 100 kilometres of the mine noted the increasing presence of AGA company representatives and their efforts to both survey and purchase land needed for the mega-project. In the neighbouring municipalities of Doima and Piedras, worries flourished around the possible impact of Anglo Gold's plans to build a metallurgical plant and a series of leaching pools. In particular, the communities wanted to avoid any impact on the quality of water in the River Opia, on which their rice crops depended. Estimates suggested that the project would require 1 cubic metre of water per second in order to process each ton of mineral, that is, the mine would use up to 31.5 million cubic metres of water per year.

Local uncertainty and lack of consultation resulted in local protests and the blocking of company traffic on roads in the area for six months starting in January 2013. The government sent in the military to protect AGA company employees entering the area (Leonor Avila, Interview, 23 September 2014). Following independent research carried out in 2013 by the Colombia Solidarity Campaign and London Mining Network, an anti-extractives NGO, it became clear to local communities and the district authorities that the company not only envisaged a single massive mine but a complex of storage areas and roads for tailings, toxic waste and refinement (i.e. the Anaima-Tocha complex). The information provided by other environmental organisations also revealed to the local population the effects of degradation, contamination and pollution caused by extractive activities on people's health and the environment (details of the ill-effects of AGA's operations in South Africa were particularly emphasised).[47] As millions of barrels of drilling fluids and formation water are dumped into rivers and forests as a result of oil production, and river courses changed and heavy metals leak into drinking water as a result of mining practices, the direct environmental costs of extraction have become all too evident and unacceptable.[48] Whether a mine is underground or open-pit, most of what is mined is discarded, leaving millions of tons of waste rock and tailings loaded with dangerous heavy metals that had previously been more or less safely bound up in the rock.

Whilst negotiations, consultations and protests followed it became increasingly evident to protesters that other tactics had to be tested. In July 2013 the population of Piedras, in the Colombian district of Tolima, cast their votes in a local referendum (*consulta popular*) to clarify levels of community opposition to the construction of an opencast gold mine named *La Colosa*. New research demonstrates the referenda in Colombia to be part of a new wave of environmental action taking place throughout Latin America.[49] The results of the popular consultation were overwhelming: from the total number of 5105 registered voters, 2971 voted against and only 24 in favour, showing a clear rejection of the population towards AGA's operations. Whilst taking place in a small rural municipality of little more than 6000 inhabitants and drawing on recognised municipal legal ordinance, the national and international media quickly recognised the Piedras referendum as an exceptional event that would inspire others. Within weeks civil society organisations in the municipality of Casanare immediately launched a proposal to carry out a referendum in the style of Piedras against oil extraction. In Tauramena, drawing on the same experience, a consultation was organised to confront exploratory activities by Ecopetrol. Within the region of Tolima, the publically perceived success of the referendum in Piedras also led directly to further calls for further referenda. Another three popular consultations are planned for 2016 (in Cajamarca, Espinal and Ibague respectively). Importantly, the regional governor as well as national politicians such as Iván Cepeda Castro, a renowned senator linked to the *Polo Democratico Alternativa* political party,[50] also granted the campaign higher-level political support.

The actions of the community of Piedras and Tolima set a new precedent for democratic participation in Colombia. It made clear a claim that the local community, represented by the municipality, represents a qualified and legitimate voice that can question the leadership of the national government. However, whilst it was clear civil society received an initial boost from the Piedras referendum, AGA and the national presidency were quick to respond. AGA publically expressed 'disagreement with the biased question'.[51] In an interview with the electronic newspaper *La Silla Vacia* in July 2014, AGA's representative, Felipe Marquez Robledo, condemned the referendum as a 'referendum of the deaf' because of the failure of campaigners to discuss and include the company in its organisation, or to guarantee truly democratic participation. Within the government more serious questions about the legality of the referendum and plans for others also started to be asked. This has resulted in an unparalleled legal confrontation between the Prosecutor General's office on one side, and the Constitutional Court and the National Office of the Controller General on the other. Both sides have made claims to being the holders of the legal and political truth in the country. Five months after the Piedras ballot and a few days following the ballot in Tauramena, President Santos stated in an interview with the national newspaper *El Espectador* that 'popular consultations ... are illegal and have no legal effect. The subsoil belongs to all Colombians. There is no room for discussion.'[52]

On the one hand, the Constitutional Court and the Council of State assert that it is necessary to consult municipal authorities in any decision regarding the use of land that could affect the environment.[53] This position is based on what they state is the constitutionally founded right of every municipality in Colombia to autonomously decide on how to best preserve the environment in its territories and to regulate the use of land appropriately. According to Chapter 4 of Law 134 (1994), which regulates the mechanisms for citizens' participation, popular consultation is a mechanism for citizen participation in which the people have the right to decide on questions of vital importance. Popular consultation may be national, departmental, municipal, district or local. There is no need to go to Congress to gain formal backing, but regional governors or local mayors must meet certain requirements that are defined in the general regulations for territorial organisation. For the popular consultation to be legal, the ballot with which the citizen exercises their vote must be merely a yes or no question. The national Senate must also carry out a supporting vote in the days following approval of the consultation. Finally, to ensure a concrete outcome, the authorities entitled to execute the query have three months to start its application. If this does not happen, the Senate is required to take steps so that the popular decision becomes binding. However, in the event that this does not take place, the president, regional governor or mayor must impose it using the force of law within a period not exceeding three months.

In contrast, the interpretation of the law offered by the office of the Prosecutor General (*Procuraduría General de la Nación*) and the executive power represented by the president and the National Mining Agency asserts that every decision regarding the use of the sub-soil and non-renewable natural resources lies within the jurisdiction of the national government. Seen from their position, no local authority is competent to establish the management and use of such resources, and therefore any popular consultation on the matter is illegal. The office of the Prosecutor General bases its position on the Mining Code (Law 685) of 2001, article 37 of which stipulates that that no regional authority, sectional or local, can exclude areas of the territory from mining either temporarily, or permanently. Regulating decree 0935/2013 further states that according to the provisions of the constitution, the state owns the sub-soil and all non-renewable natural resources (articles 5, 7 and 10 of Law 685/2001). As such, the state has the obligation to conserve these goods as well as the competence and power to grant special rights to use these resources through concessions. These constitute a special legal power for securing the public good (Decision C-983/10).

Given their interpretation, the Prosecutor General's office has proceeded with an official investigation of the mayor and Municipal Council of Piedras. This in effect freezes the accounts of the local municipality and places all further administrative decision-making by the office in legal question. They have also released a formal warning to all the municipal officials that following investigation a series of disciplinary measures may be taken. The Prosecutor General's office has furthermore publicly warned other municipalities in the region contemplating a referendum on the mine. They will take similar action against them in their event, and this has had some impact. The Municipal Council of Cajamarca decided in February 2015 (10 votes to 1) that they could not support a decision calling for referenda on the mine project, as this was not within their competence. This decision made direct reference to the recommendations of the Prosecutor General's office. The council also made clear reference to the pressure placed on it by AGA. As leaks to the press revealed, AGA had sent a letter to the municipal council stating that all social and economic programmes started in the community during the exploration stage would stop if the local municipality carried out a popular consultation.[54]

In an attempt to counter what they now see as a witch-hunt being waged by the Prosecutor General's office, further steps have been proposed by the Constitutional Court and Council of State to support the legality of municipal referenda on extractive activities. They have suspended decision C-123/2014 of the Constitutional Court which declares that the competent national authorities will agree with the concerned local authorities all measures necessary for healthy environmental protection, in particular watersheds, the economic, social and cultural development of communities and the health of the population. To do this they applied article 288 of the constitution, which outlines principles for coordination, competition and subsidiarity based on the autonomy of local authorities within their territories. They have furthermore attempted to suspend decree 935/2013 that states what the applicable requirements are to obtain concessions as well as the financial requirements applicants for new concession contracts must meet. The Constitutional Court has suspended the decree because of what it perceives as an unnecessary addition to the requirements needed to obtain a mining licence. The government did not contemplate these changes during the initial formation of Law 685/2001.

Attempting to respond to the suspension of decree 935/2013 by the Constitutional Court and the Council of State, the national presidency countered by issuing decree 2691/2014. This is an attempt to limit the legal power under article 37 of the Mining Code, whereby local authorities can require measures necessary for the protection of the environment. In order to achieve this the new decree asks local authorities to finance technical studies (EIAs) that show the possible impact of mining projects on the environment. The national presidency argues that this will speed up the provision of scientific criteria on which to base an educated judgment on whether to allow or stop mining activity. However, as critics point out, the majority of local governments do not have the resources to conduct such studies, and it is still not possible for a local authority to decide on its own to exclude a part of its territory from extractive activities.[55] The power of decision-making therefore remains with the central government.

**Targeted violence**

The current legal battle between different branches of government and its impact on the processes of local government and civil society participation are a form of structural violence to the political-legal order. The displacement of people and environmental impacts of the Colombian extractive economy are direct forms of violence. The direct physical violence associated with extraction cannot simply be excused by the conditions of the ongoing armed conflict. The nation's murder rate is at a 30-year low and foreign investment is up, yet Colombia remains the most dangerous place in the world to be a trade unionist. Every year numerous union leaders, union activists and union members are assassinated – over 250 in the past 20 years.[56] Organisations working to support workers' rights in Colombia report 22 unionists were killed in the country in 2015. Some of these cases are directly linked to union action in connection with extractive activities, including the national oil company Ecopetrol.[57] Further civil lawsuits have also been filed against US companies including Coca Cola, Dole and Drummond, for allegedly using paramilitaries to kill trade unionists.[58]

Targeted assassination of human rights and environmental activists also continues at a shocking pace in Colombia. The extremely low number of prosecutions also makes it clear that these killings take place with a high degree of impunity with regard to state authorities. The International Office of Human Rights – Action Colombia (OIDHAC) released a statement in November 2015 in which they said:

> Since 1997, the office has been following the attacks and aggression that affect human rights defenders. In the work together with the National Prosecutors Office, civil society we have succeeded in putting together a list of 729 homicides of human rights defenders that occurred between 1994 and 2015, indicating an annual average of 33 assassinations. Almost all of these cases remain in impunity. In the first nine months of this year, the office registered 30 homicides and 20 cases of attempted homicide against defenders of human rights. This demonstrates the persistence of internal insecurity and hostility towards the exercise of their work. With these recent cases, this year has surpassed the annual average for homicides of human rights defenders registered over the last 20 years (OIDHAC 2015-11-19, Bogotá, DC).[59]

Again some of these cases can be attributed to aggressive reactions towards those opposing extractive activities in the country. In the case of Tolima described above, the assassination of Pedro César García Moreno a member of the peasant farmer organisation *Consiencia Campesina*, who had been active in opposing the mine project, was a key factor in kick-starting the referendum and establishment of a regional social movement.[60] Similar killings continue to be reported by the Colombian press and civil society organisations.[61] During my field research and contact with people in human rights organisations and the leadership of the social movement against the mine in Tolima, that is, the Committee for the Defense of Life, I was repeatedly informed of the messages, signed by the names of locally active Bandas Criminales (BACRIM) i.e. criminal bands or paramilitary groups, they had received on their private phones warning that if they did not desist from their activities they would be killed. A recent Global Witness report draws attention to the same threats and the 25 environmental activists that were killed in Colombia in 2014.[62]

## Conclusion

The claim that the continuance of the extractive economy is a necessary driver for economic development, and source of financing for the post-conflict period, needs to be contrasted with the consequences of this decision for human rights and the environment. As this article demonstrates, even if the peace accords with the FARC are ratified through a national public referendum, much of the violence linked to the mining and oil and gas extraction in the country is likely to continue. Whilst the peace accords will be an important advance towards the establishment of peace and sustainable development in the country, they will not in their own right be enough to end the insecurities generated directly and indirectly by the extractive economy. As I have highlighted, in recent years there has been a re-articulation, in line with Reno's idea of the 'shadow state',[63] of the armed conflict and of the warring parties involved with the capture and control of extractive resources. Whilst the government claims it is responding harshly to illegal mining, it continues to take little public action against armed and criminal groups who are linked to large-scale legal mining.

Moreover, as I have attempted to demonstrate in this article, the steps taken in recent years by the government to ease and simplify the environmental licensing process, indicate that the Colombian state is deliberately flouting constitutionally founded principles support-ing popular sovereignty and local democratic governance of the local environment. Efforts to innovatively use existing democratic principles, mechanisms and regulations (from protest to class action lawsuits, prior consultation and popular referenda) are met with state repression, threats and legal action, countering the expression of human rights. More desperate still, activists involved in these efforts to defend territory and the environ-ment continue to find themselves at threat of direct violence and assassination. Rather than a means to extract justice, or to pay for peace as the Colombian government claims, the con-tinued commitment to the expansion of extractive activities and the mutating state of inse-curity in the country will move the Colombia into a new phase of conflict. Simply put, all indications reinforce the analysis that justice will never be 'extracted'.

## Disclosure statement

No potential conflict of interest was reported by the author.

## Funding

This work was supported by the Norwegian Research Council [Lat-Am 236912].

## Notes

1. Colombian National Development Plan, https://www.dnp.gov.co/Plan-Nacional-de-Desarrollo/PND-2010-2014/Paginas/Plan-Nacional-De-2010-2014.aspx
2. Extracting Justice Project, http://www.nmbu.no/en/aboutnmbu/faculties/samvit/departments/noragric/research/clusters/rapid/projects-and-assignments/extracting-justice
3. Micheal Ross, *The Oil Curse: How Petroleum Wealth Shapes the Development of Nations* (Prin-ceton, NJ: Princeton University Press, 2013).
4. Terry Lynn Karl, *The Paradox of Plenty: Oil Booms and Petro States* (Berkeley: University of California Press, 1996).
5. Over five million people. See: UNHCR, 'Worldwide Displacement Hits All-time High as War and Persecution Increase', 18 June 2015, http://www.unhcr.org/558193896.html
6. See P. Collier and A. Hoeffler, 'Aid, Policy and Growth in Post-War Countries', *European Economic Review* 48 (2004): 1125–45; Humphrey Macartan, Jeffrey Sachs, and Joseph Stiglitz, *Escaping the Resource Curse* (New York: Columbia University Press, 2007).
7. See Phillipe Le Billon, 'The Political Ecology of War: Natural Resources and Armed Conflicts', *Political Geography* 20 (2001): 561–84; John-Andrew McNeish and Owen Logan, eds, *Flam-mable Societies: Studies on the Socio-Economics of Oil and Gas* (London: Pluto Press, 2012); Timothy Mitchell, *Carbon Democracy: Political Power in the Age of Oil* (New York: Verso Books, 2011); Hannah Appel, Arthur Mason, and Micheal Watts, *Subterranean Estates: Life Worlds of Oil and Gas* (Ithaca, NY: Cornell University Press, 2015).
8. FARC announced the cessation of all attacks on 9 February 2015. The ceasefire was nonetheless broken over several weeks in the spring of the same year.
9. CERAC Monthly Report on the De-escalation of the Armed Internal Conflict in Colombia, http://blog.cerac.org.co/segundo-reporte-de-monitoreo-mensual-de-medidas-de-desescalamiento-del-conflicto-armado-interno-en-colombia

10. 'Ex-president Uribe Wages One-man Twitter War against Colombia Peace Deal', *Washington Post*, 25 September 2015, https://www.washingtonpost.com/news/worldviews/wp/2015/09/25/ex-president-uribe-wages-one-man-twitter-war-against-colombia-peace-deal/

11. '¿Es suficiente el cese al fuego bilateral para lograr la paz en Colombia?' *Telesur*, 29 October 2015, http://www.telesurtv.net/news/Con-el-cese-al-fuego-bilateral-se-logra-la-paz-en-Colombia-20151028-0044.html

12. Dejusticia, http://www.dejusticia.org/-!/index

13. César Rodríguez Garavito, '¿Paz territorial sin paz ambiental?' *El Espectador* 2 December 2015, http://elespectador.com/print/543784

14. Ibid.

15. Fabrizio Hoschield, *Consideraciones Ambientales Para la Construcción de Paz: Territorial Estable, Duradera y Sostenible en Colombia* (Bogota: PNUD, 2014).

16. Ted Dunning and Leslie Wirpsa, 'Oil and the Political Economy of Conflict in Colombia and Beyond: A Linkages Approach', *Geopolitics* 9, no. 1 (2004): 88–108; Leonardo Gonzalez Parafán, ed., *Impacto de la Minera de Hecho en Colombia* (Colombia: Punto de Encuentro, 2013); Francisco Ramirez Cueller and Aviva Chomsky, *The Profits of Extermination: Big Mining in Colombia* (Monroe, ME: Common Courage Press, 2005).

17. Angelika Rettberg, 'Gold, Oil and the Lure of Violence: The Private Sector and Post-conflict Risks', *NOREF Report Series* (2015).

18. CODHES, 'Minería y palmicultura, nuevas causas de desplazamiento en Colombia', Pacifico-colombia.org, 22 February 2011, http://www.pacificocolombia.org/novedades/codhes-mineria-y-palmicultura-nuevas-causas-de-desplazamiento-en-colombia/382

19. 2015 UNHCR Country Operations Profile – Colombia, http://www.unhcr.org/pages/49e492ad6.html

20. COHDES Bulletin Number 79, March 2012, http://www.acnur.org/t3/uploads/media/CODHES_Informa_79_Desplazamiento_creciente_y_crisis_humanitaria_invisibilizada_Marzo_2012.pdf?view=1

21. Indepaz, 'Impacto de la minería de hecho en Colombia', Bogota, noviembre de 2012, 12.

22. Daniel Tubb, *Gold in the Chocó* (Unpublished PhD Thesis, Carleton University, Canada, 2014).

23. Raul Zabechi, 'Mining and Post-conflict in Colombia', Americas Program, 3 July 2014, http://www.cipamericas.org/archives/12465

24. Adam Isachson, 'A Bumpy Ride Ahead: Civil Military Relation Challenges Awaiting Post-Conflict Colombia' (Unpublished paper presented at LASA 2014).

25. On rural reform, demobilisation and political participation of ex-combatants, combatting the illicit drug trade, and the compensation of victims.

26. OPALC, 'La Tierra: Desarrollo Agrario y el Sector Extractivo', http://www.sciencespo.fr/opalc/sites/sciencespo.fr.opalc/files/Incentivoseconómicos.pdf

27. The other drivers included agriculture, housing, innovation and infrastructure. See Colombian National Development Plan 2010–2014, https://colaboracion.dnp.gov.co/CDT/PND/PND2010-2014TomoICD.pdf

28. Juan Manuel Santos, *Report to the Congress 2014*, http://wsp.presidencia.gov.co/Publicaciones/Documents/InformePresidente2014.pdf

29. 'Financing Peace: The Colombian Economy after the FARC', Global Risk Insights, 12 October 2015, http://globalriskinsights.com/2015/10/financing-peace-the-colombian-economy-after-the-farc/

30. 'Colombia Seeks Dividend from the Wealthiest', *Financial Times*, 17 September 2014, http://www.ft.com/intl/cms/s/56440c30-3d29-11e4-a2ab-00144feabdc0,Authorised=false.html?siteedition=intl&_i_location=http%3A%2F%2Fwww.ft.com%2Fcms%2Fs%2F0%2F56440c30-3d29-11e4-a2ab-00144feabdc0.html%3Fsiteedition=intl&_i_referer=http%3A%2F%2Fglobalriski

31. Ibid.

32. Reserves have not been growing at anything like the speed of production. Colombia currently has just seven years of proven reserves. In 2014, the government ran a 2.1% deficit (a gap of some US$5.1 billion), thus experiencing a 'twin-deficit' from its fiscal and trade balances. Expectations of a rise in US interest rates and plunging commodity prices (coffee by 31.2%, coal 32.6%, nickel 43% and gold by 16.2%) prompted a drastic devaluation of Colombia's currency by nearly 60% over the year 2015–2016.

33. 'Costo-Beneficio del decreto de las licencias express', *La Republica*, 11 February 2015, http://www.larepublica.co/costo-beneficio-del-decreto-de-las-licencias-express_218816
34. Ibid.
35. Colombia National Constitution 1991, https://www.constituteproject.org/constitution/Colombia_2005.pdf
36. According to the latest national census (2005) 3.4% of the Colombian population is indigenous, and 10.4% is of afro-descendent background.
37. *The Right of Indigenous Peoples to Prior Consultation: The Situation in Bolivia, Colombia, Ecuador, and Peru*, Oxfam Report, 11 April 2011, http://www.oxfamamerica.org/explore/research-publications/the-right-of-indigenous-peoples-to-prior-consultation-the-situation-in-bolivia-colombia-ecuador-and-peru/
38. Report of the Special Rapporteur on the Situation of Human Rights and Fundamental Freedoms of Indigenous Peoples, Mr James Anaya. The Situation of Indigenous Peoples in Colombia: Follow-up to the Recommendations Made by the Previous Special Rapporteur, 25 May 2010, http://www2.ohchr.org/english/bodies/hrcouncil/docs/15session/A.HRC.15.37.Add.3_en.pdf
39. Diana Maria Ocampo and Sebastian Agudelo, 'Case Study: Colombia', *Americas Quarterly*, Special Edition Consulta Previa and Investment (Spring 2014), http://americasquarterly.org/content/consulta-previa-and-investment
40. Reprinted in Resumen, Diálogos interculturales sobre minería en Colombia. Propuestas desde el Resguardo Indígena Cañamomo Lomapreita y el Palenque Alto Cauca, 16–17 April 2015, Santender de Quilichao, Cauca, Forest Peoples Programme/PCN/Embassy of Switzerland in Colombia.
41. Cesar Rodriguez Gavarito, El Espectador, 29 April 2013, http://www.elespectador.com/opinion/locomotora-sin-rieles
42. Deborah Poole, 'The Minga of Resistance: Policy-Making from Below', *NACLA* (2008), https://nacla.org/news/minga-resistance-policy-making-below
43. 'Fogonazo y memoria del Paro', *Desde Abajo*, 23 September 2013, http://www.desdeabajo.info/ediciones/item/22777-fogonazo-y-memoria-del-paro.html
44. Anglo Gold Ashanti (AGA) is the third-largest gold-mining multinational in the world, with a presence in various countries including South Africa, Tanzania, Ghana, Congo and Colombia.
45. Anglo Gold Ashanti Annual Report 2010, http://www.anglogoldashanti.com/en/Media/Reports/AnnualReports/AGA-annual-review-2010.pdf
46. Anglo Gold Ashanti Annual Report 2012, http://www.anglogoldashanti.com/en/Media/Reports/AnnualReports/AGA-mineral-resource-and-ore-reserve-report-2012.pdf
47. For example, 'Anglo Gold Ashanti Faces Charges for Polluting South Africa's Vaal River', Mining.com, 2 September 2013, http://www.mining.com/anglogold-ashanti-faces-charges-for-polluting-south-africas-vaal-river-66700/
48. James Kneen, 'Earth: What is Mining All About? The Up and Down Sides' (Part of the Ottawa University Institute of Environment 2006–07 Lecture Series 'Elemental Environmentalism: Water, Air, Fire and Earth', 28 February 2007), http://miningwatch.ca/sites/default/files/Mining_Unsustainable_0.pdf
49. Stuart Kirsch, *Mining Capitalism: The Relationship between Corporations and their Critics* (Durham, NC: Duke University Press, 2014).
50. Cepeda Castro is well-known in Colombia as a representative for the Movement of Victims of the Crimes of the State (MOVICE) and for his public accusation of the existence of clear connections between ex-President Uribe and the human rights abuses of paramilitary organisations.
51. The text of the ballot read: 'Do you agree, as a resident of Piedras, Tolima that in our jurisdiction, the following activities are carried out: exploration, exploitation, treatment, transformation, transportation, washing of materials that originate from large scale gold mining; or that materials that are harmful to health and the environment are stored or used, specifically cyanide and/or any other substances or hazardous materials associated with these activities; furthermore, that surface and ground water is used from our town in such operations or any other similar operations that may affect and/or limit the supply of potable water for human consumption, and for agriculture, the traditional productive vocation of our municipality?'
52. 'Se puede ganar en primera vuelta': Santos, *El Espectador*, 21 December 2013, http://www.elespectador.com/noticias/politica/se-puede-ganar-primera-vuelta-santos-articulo-465498

53. Decision C-123/201, Constitutional Court, Auto 11001032600020130009100 (47693), Feb. 26/ 14, C.P. Jaime Orlando Santofimio, Council of State.

54. El Tiempo, 'Concejo de Cajamarca dice no a la consulta popular minera', http://www.eltiempo. com/colombia/otras-ciudades/consulta-popular-minera-en-cajamarca/15247615 (accessed 15 April 2015).

55. 'El regalo minero de Santos a los alcaldes terminó siendo de arena', *La Silla Vacía* 13 March 2015, http://lasillavacia.com/historia/el-regalo-minero-de-santos-los-alcaldes-termino-siendo-de-arena-49743

56. 'End Anti-Trade Union Violence in Colombia', Justice for Colombia, http://www. justiceforcolombia.org/campaigns/union-rights/

57. 'Colombia One of the Terrible Ten Worst Countries for Workers Rights', *Stronger Unions*, 16 June 2015, http://strongerunions.org/2015/06/16/colombia-one-of-the-terrible-ten-worst-countries-for-workers-rights-3/

58. Adam Roston, 'It's the Real Thing, Murder. US Employers Like Coca-Cola Are Implicated in Colombia's Brutality', *The Nation*, 23 August 2001, http://www.thenation.com/article/its-real-thing-murder/

59. My translation: Desde 1997, la Oficina viene haciendo un seguimiento a los ataques y agresiones que afectan a los defensores de derechos humanos. En un trabajo conjunto entre la Fiscalía General de la Nación, las organizaciones de la sociedad civil y la Oficina, se logró consolidar un listado de 729 homicidios de defensores de derechos humanos, ocurridos entre 1994 y 2015, lo que indica un promedio anual de 33 asesinatos. Casi todos estos hechos permanecen en la impunidad. En los nueve primeros meses del año, la Oficina registró 30 homicidios y 20 tentativas de homicidio contra defensores de derechos humanos, lo que demuestra la persistencia de un entorno inseguro y hostil para el ejercicio de su labor. Con estos casos recientes, este año se ha superado el promedio de homicidios de defensores de derechos humanos que se venía registrando en los últimos 20 años.

60. 'Environmentalist Opposed to Central Colombia Mining Project Assassinated', *Colombia Reports*, 5 November 2013, http://colombiareports.com/farmers-leader-environmental-acitivist-assassinated-near-colombian-capital-bogota/

61. 'Ya son tres lideres campesinos asesinados en una semana: presumen motivos politicos', Colombiainforma.info, 20 November 2015, http://colombiainforma.info/mov-sociales/ pueblos/2899-ya-son-tres-los-lideres-campesinos-asesinados-en-una-semana-presumen-motivos-politicos

62. 'Report: Latin America Most Dangerous Region for Land Activists', *Aljazeera*, 20 April 2015, http://america.aljazeera.com/articles/2015/4/20/report-latin-america-most-dangerous-for-land-activists.html

63. William Reno, 'Clandestine Economies, Violence and States in Africa', *Journal of International Affairs* 53, no. 2 (2000).

# Index

Printed and bound by CPI Group (UK) Ltd, Croydon, CR0 4YY

01/11/2024

01782600-0004